NANOTECHNOLOGY SCIENCE AND TECHNOLOGY

TITANIUM DIOXIDE

ADVANCES IN RESEARCH AND APPLICATIONS

NANOTECHNOLOGY SCIENCE AND TECHNOLOGY

Additional books and e-books in this series can be found on Nova's website under the Series tab.

NANOTECHNOLOGY SCIENCE AND TECHNOLOGY

TITANIUM DIOXIDE

ADVANCES IN RESEARCH AND APPLICATIONS

APARNA B. GUNJAL
EDITOR

Copyright © 2022 by Nova Science Publishers, Inc.

DOI: https://doi.org/10.52305/SUWN8331

All rights reserved. No part of this book may be reproduced, stored in a retrieval system or transmitted in any form or by any means: electronic, electrostatic, magnetic, tape, mechanical photocopying, recording or otherwise without the written permission of the Publisher.

We have partnered with Copyright Clearance Center to make it easy for you to obtain permissions to reuse content from this publication. Simply navigate to this publication's page on Nova's website and locate the "Get Permission" button below the title description. This button is linked directly to the title's permission page on copyright.com. Alternatively, you can visit copyright.com and search by title, ISBN, or ISSN.

For further questions about using the service on copyright.com, please contact:
Copyright Clearance Center
Phone: +1-(978) 750-8400   Fax: +1-(978) 750-4470   E-mail: info@copyright.com

NOTICE TO THE READER

The Publisher has taken reasonable care in the preparation of this book, but makes no expressed or implied warranty of any kind and assumes no responsibility for any errors or omissions. No liability is assumed for incidental or consequential damages in connection with or arising out of information contained in this book. The Publisher shall not be liable for any special, consequential, or exemplary damages resulting, in whole or in part, from the readers' use of, or reliance upon, this material. Any parts of this book based on government reports are so indicated and copyright is claimed for those parts to the extent applicable to compilations of such works.

Independent verification should be sought for any data, advice or recommendations contained in this book. In addition, no responsibility is assumed by the Publisher for any injury and/or damage to persons or property arising from any methods, products, instructions, ideas or otherwise contained in this publication.

This publication is designed to provide accurate and authoritative information with regard to the subject matter covered herein. It is sold with the clear understanding that the Publisher is not engaged in rendering legal or any other professional services. If legal or any other expert assistance is required, the services of a competent person should be sought. FROM A DECLARATION OF PARTICIPANTS JOINTLY ADOPTED BY A COMMITTEE OF THE AMERICAN BAR ASSOCIATION AND A COMMITTEE OF PUBLISHERS.

Additional color graphics may be available in the e-book version of this book.

Library of Congress Cataloging-in-Publication Data

ISBN: 978-1-68507-457-9

Published by Nova Science Publishers, Inc. † New York

Contents

Preface vii

Chapter 1 **Applications of Titanium Dioxide in Space** 1
Naoki Shimosako and Hiroshi Sakama

Chapter 2 **Revolution of Titanium Dioxide
in Biomedical and Applications
in Environmental Remediation**................................ 17
Smita Kumari and Dharmendra Kumar

Chapter 3 **Doped Titanium Dioxide Nanostructures
for Visible Light-Driven Photocatalysis**................... 49
Seema Maheshwari, Kuldeep Kaur,
Ashok Kumar Malik and Simrat Kaur

Chapter 4 **Biosynthesized Titanium Dioxide
Nanoparticles and Their
Antibacterial Efficacy**.. 73
Immanuel J. Suresh, Iswareya V. Lakshimi
and Shanthipriya Ajmera

Chapter 5 **Forensic Applications of
Titanium Dioxide Nanomaterials**........................... 103
Gurvinder Singh Bumbrah
and Devidas S. Bhagat

About the Editor... 121

Index 123

PREFACE

This book, titled *Titanium dioxide: Advances in Research and Applications*, deals with important aspects of applications of titanium dioxide and recent developments in the areas of research of titanium dioxide. There are 5 chapters in this book contributed by various authors.

Chapter 1 deals with applications of titanium dioxide in space and its photocatalytic activity in a vacuum environment.

Chapter 2 mentions various applications of titanium dioxide in the biomedical field, such as in bone implantation, immobilization of drug carriers, and in cancer therapy as well as in environmental remediation, such as degradation of hazardous pollutants and treatment of wastewater.

Chapter 3 deals with doped titanium dioxide nanostructures for visible light driven photo catalysis. The applications of visible light on active titanium dioxide photocatalysts for degradation of various toxic pollutants are described.

Chapter 4 deals in detail with various methods of biosynthesis of titanium dioxide nanoparticles and their antibacterial activity.

Chapter 5 describes preparation of titanium dioxide nonmaterials, their forensic applications, and fingerprint imaging.

The book *Titanium dioxide: Advances in Research and Applications* will be of immense use to students, professors and researchers of various colleges, universities and research institutes. It will increase knowledge regarding the synthesis, applications and various aspects of titanium dioxide.

In: Titanium Dioxide
Editor: Aparna B. Gunjal
ISBN: 978-1-68507-457-9
© 2022 Nova Science Publishers, Inc.

Chapter 1

APPLICATIONS OF TITANIUM DIOXIDE IN SPACE

Naoki Shimosako[1] and Hiroshi Sakama[1,*]
[1]Department of Engineering and Applied Sciences,
Sophia University, Kioi-cho, Chiyoda-ku,
Tokyo, Japan

ABSTRACT

We have been studying the application of titanium dioxide (TiO_2) to spacecraft materials. Materials for spacecraft must be durable in the space environment. We introduce the durability of TiO_2 under electron beam and atomic oxygen irradiation. In addition, the photocatalytic activity of TiO_2 in orbit has been evaluated as a solution to the contamination problem of spacecraft. We also introduce the contamination problem faced by spacecraft and the effect of the vacuum environment on photocatalytic activity.

Keywords: TiO_2, Space, Electron beam, Atomic oxygen, Photocatalyst, Contamination, Vacuum

---

[*] Corresponding Author's E-mail: h-sakama@sophia.ac.jp.

INTRODUCTION

Titanium dioxide (TiO_2) is used in various applications, including paints, photocatalysts, charge-transport layers in solar cells, and optical elements. We have been studying the application of TiO_2 in materials for spacecraft. Because the space environment is harsh, spacecraft materials must be highly durable.

Here we describe the effects of radiation and atomic oxygen (AO), which are encountered in the space environment and can degrade materials, and investigate the ability of TiO_2 to mitigate such degradation. In addition, the photocatalytic function of TiO_2 is found to be effective for decontamination not only on the ground but also in space. We also discuss the contamination problems related to spacecraft and the effect of the vacuum environment on photocatalytic activity.

RADIATION

Three types of radioactive environments affect spacecraft: the Van Allen radiation belts, galactic cosmic rays, and solar cosmic rays [1]. For a spacecraft orbiting the Earth, the most important of these are the Van Allen radiation belts. Galactic cosmic rays have high energy, but the dose is negligibly small.

Solar cosmic rays are the primary radiation environment in interstellar space, but not in Earth's orbit. The Van Allen radiation belts are regions of energetic radiation particles trapped by the Earth's magnetic field, and form doughnut shapes along the equator. These regions are present from low-altitude orbit to high-altitude orbit. They consist mainly of electrons with energies as high as several megaelectron volts and protons with energies as high as several hundred megaelectron volts.

Such radiations can create defects in materials. The effects of radiation on semiconductors can be divided into two main categories: the total-dose effect and the single-event effect (SEE). In the total-dose effect, the cumulative effect of radiation damages the material. The total-dose effect can be further classified into ionization damage and displacement damage. Ionization damage is due to electron-hole pairs produced by radiation. Displacement damage is caused by displacement of nuclei from their original positions in the lattice, leading to degradation. Optical devices, charge-coupled devices, and solar cells are susceptible to displacement damage. The total-dose effect is mainly caused by trapped particle radiation by the Earth's magnetic field and solar protons. In an SEE, an electronic device malfunctions as a result of impact by a single heavy-ion particle. It is mainly caused by energetic particles in galactic cosmic radiation, trapped particle radiation by the Earth's magnetic field, and solar proton radiation. The main countermeasures against such radiation are to install radiation shields and to use radiation-resistant materials.

Table 1 summarizes the reports on the irradiation of TiO_2 with electron beams (EBs).

Table 1. Reports on the irradiation of TiO_2 with EBs

Sample	Irradiation energy (MeV)	Absorbed dose (kGy)	Atmosphere	Ref.
TiO_2/cyanate ester	0.03–0.2	–	Vacuum	2
TiO_2	0.2	12	Air	3
TiO_2	1	500–5000	He	4
TiO_2	1	5–15	Air	5
TiO_2/polyacrylonitrile	1	10000	–	6
TiO_2/carbon fiber	1	5	Air	7
TiO_2	1	500–2000	Air	8
TiO_2	1.14	0–100	Ar	9
N-doped TiO_2	2	140–500	–	10
F-doped TiO_2	6	30–980	–	11

Qin et al. evaluated the EB resistance of TiO_2 for space applications [2]. They prepared TiO_2/cyanates and reported that TiO_2 increased the resistance of cyanates to EB radiation. Other reasons for studying EB radiation include improving photocatalytic activity. Such studies typically aim to develop TiO_2 with high photocatalytic activity by generating defect levels via EB irradiation. For example, Jun et al. irradiated TiO_2 with EBs in helium atmosphere and compared its photocatalytic activity before and after the irradiation [4]. They found that oxygen vacancies and Ti^{3+} states were generated on the surface of TiO_2 and that these defect levels reduced the rate of electron-hole recombination, thereby enhancing the photocatalytic activity. However, they also found that the photocatalytic activity decreased with increasing radiation dose. The photocatalytic activity is determined by the competition between the charge separation effect at defect levels and nonradiative recombination. TiO_2 has been reported to absorb in the visible-light region because of defect levels, resulting in visible-light-responsive photocatalysts. Although enhancing the photocatalytic activity by absorbing visible light is a desirable effect, the transformation of TiO_2 into a visible-light-absorbing material in space makes it unsuitable as a surface material for spacecraft. The thermo-optical properties of surface materials for spacecraft have been rigorously evaluated.

The heat increase due to light absorption by the spacecraft was calculated in advance, and the cooling effect of radiative emission was determined in order to balance it. When EB irradiation converts TiO_2 into a visible-light-absorbing material in space, the spacecraft temperature may increase because the amount of cooling is insufficient. Therefore, ideally, the thermo-optical properties of TiO_2 should not be altered by radiation exposure.

Our group has evaluated the resistance of TiO_2 to EB irradiation [12]. Figure 1 shows Ti2p and O1s the X-ray photoelectron spectroscopy (XPS) spectra of TiO_2 before and after EB irradiation. The Ti2p and O1s spectra did not change after the sample was irradiated with the EBs. The other properties of TiO_2, such as its transmittance, crystallinity, and photocatalytic activity, were also unchanged.

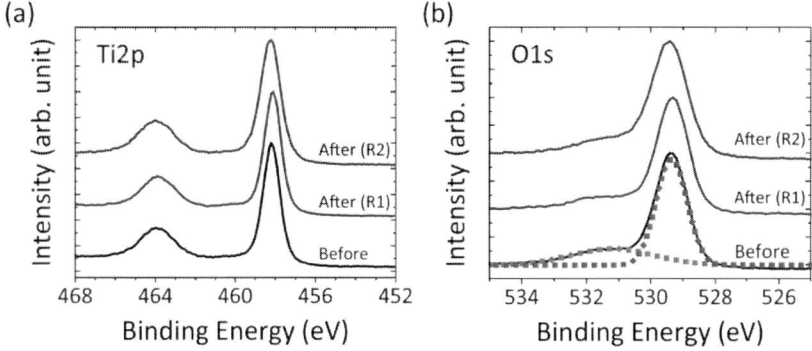

Figure 1. Typical Ti2p and O1s XPS spectra of TiO$_2$ after and before EB irradiation [12]. R1 and R2 indicate the different irradiation energies of EBs, 120–140 keV and 460–480 keV, respectively.

These results indicate that TiO$_2$ is sufficiently resistant to EB irradiation.

In addition to the effects of electrons, those of protons should also be investigated. There have been some reports that the photocatalytic activity of TiO$_2$ can be improved by irradiating it with protons to generate defects in the TiO$_2$ lattice [13, 14]. In addition, Di et al. modified the silicone rubber with TiO$_2$ powder to enhance its resistance to protons and enable its use in space applications [15]. These irradiation energies are tens to hundreds of kiloelectron volts, and further experiments with higher energies on the order of megaelectron volts are required for space applications.

ATOMIC OXYGEN

AO is produced by the dissociation of O$_2$ in the upper atmosphere by solar ultraviolet light, and is known to degrade spacecraft materials [1, 16-19]. At altitudes of 200 km to 600 km, AO is the major component of the atmosphere. At an altitude of 400 km, which corresponds to the orbit of the International Space Station, the AO density is approximately 10^8 atoms/cm^2. A spacecraft in low earth orbit (LEO), which is moving at an orbital velocity of ~8 km/s, will collide with AO at relative energy of ~5 eV. Therefore, the AO flux impinging a spacecraft flying at an altitude of 400 km is ~8 × 10^{13} atoms/s/cm^2.

AO is known to oxidize spacecraft materials. When organic materials are oxidized, the resultant oxides are almost always volatile; thus, the organic materials are eroded. For example, polyimide films are frequently used in spacecraft because of their high resistance to heat and radiation; however, such materials have been reported to disappear in orbit because of the formation of volatile substances such as CO_x under AO irradiation [16].

Numerous studies have been conducted to improve the resistance of organic materials to AO. The three main countermeasures are coating organic materials with an inorganic oxide layer [20–27], fabricating inorganic/organic hybrid materials [28–31], and preparing self-organized films [32, 33]. An inorganic oxide coating can protect an organic material. In particular, SiO_2 [20–23], Al_2O_3 [20, 23, 24], and indium tin oxide [25, 26] have been used as coating materials. The inorganic/organic hybrid materials can reduce the erosion rate. AO erodes only the organic material exposed on the surface of the inorganic/organic hybrid material.

After this erosion occurs, the inorganic material that exists deep inside is exposed and the inorganic material protects the organic material inside, stopping erosion. In self-organized films, the surface of the organic material is oxidized by AO irradiation and the oxide film protects the underlying organic material. For instance, for polysiloxane–polyimide films, the polysiloxane is oxidized by AO irradiation to produce a SiO_2 layer, which subsequently protects the material against further oxidation [32].

Because TiO_2 is a well-known inorganic oxide, it has been reported as a protective material for organic materials against AO [24, 28, 34, 35]. Xiao et al. and Tsai et al. have reported that the resistance to AO and the thermal stability of a polymer can be enhanced by hybridizing the polymer with TiO_2 [28, 35]. Gouzman et al. reported that the resistance to AO of a polyimide film was improved by coating the film with TiO_2 [34], and Minton et al. reported that a TiO_2 coating on organic materials prevented degradation by vacuum ultraviolet ray in addition to AO degradation [24]. Thus, TiO_2 is expected to function as a protective material against AO.

We have investigated the degradation of the basic properties, including the photocatalytic activity, of pure TiO_2 irradiated with AO [36]. As shown in Figure 2, the surface of TiO_2 was oxidized by AO irradiation but its transmittance and crystal structure were almost unchanged, and its photocatalytic activity was maintained at ~50%. Thus, pure TiO_2 was found to be resistant to AO. However, because TiO_2 also maintained its photocatalytic activity after AO irradiation, it should not be used to protect organic materials because the photocatalytic function of TiO_2 may degrade the organic materials it is intended to protect.

PHOTOCATALYSIS AND CONTAMINATION OF SPACECRAFT

As previously mentioned, the photocatalytic function of TiO_2 can degrade the organic materials it is intended to protect. However, a photocatalyst can also solve the problem of contamination of spacecraft in orbit. Examples of the observation wavelengths of spacecraft are summarized in Figure 3 [37-45].

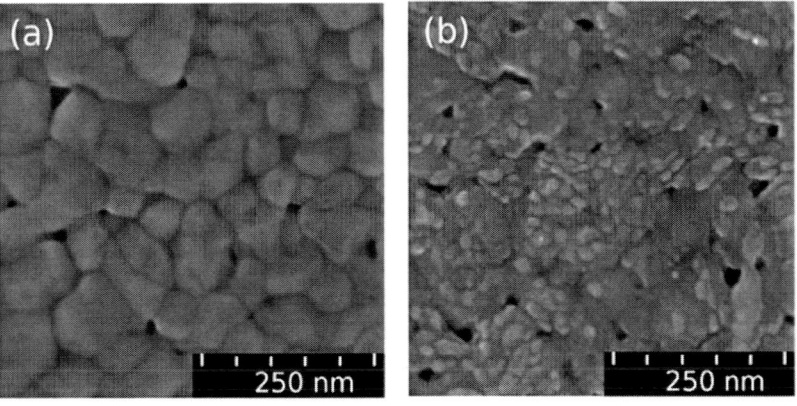

Figure 2. Surface scanning electron microscope images of TiO_2 (a) before and (b) after AO irradiation [36].

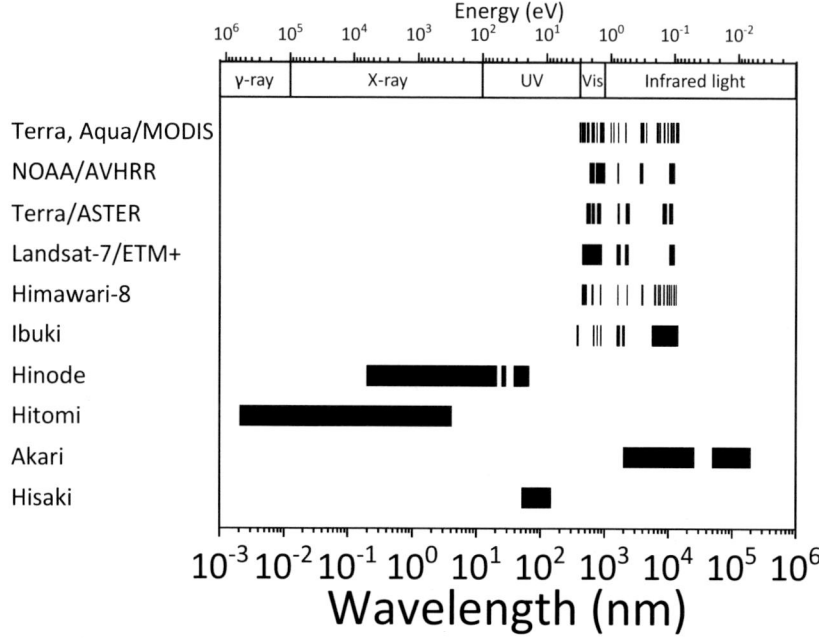

Figure 3. Examples of the observation wavelengths of spacecraft [37-45].

The wide range of wavelengths from gamma-ray to infrared ray was observed by spacecraft. The data obtained from spacecraft are important for both academic and commercial purposes. However, the intensities measured using instruments onboard spacecraft have been reported to decrease with time. The measured values decreased by as much as 20% in 10 months for the Earth observation satellite Midori-2 [46] and by as much as 70% in 4 years for the solar observation satellite Hinode [46, 49]. In addition, the solar observation satellite OSO-8 could not continue its mission because of reflectivity loss after arriving in orbit [49, 50]. It has been concluded that the decrease in signal intensity was caused by adhesion of contaminants outgassed from the organic materials in the spacecraft, which absorbed and scattered light, adversely affecting the observation results [19, 51, 52]. Because spacecraft materials must be lightweight, many organic materials such as cable sheathing, adhesives, and paints are used. Additives and unreacted substances in these organic materials volatilize when the spacecraft is in orbit. The effects of these contaminants on spacecraft have been recognized since the 1960s; however, in recent years, the increased sensitivity of sensors onboard satellites and the increased use of composite materials such as carbon-fiber-reinforced plastics to reduce the weight of satellites have led to an increase in the amount of outgassed contaminants. As a result, contamination control for spacecrafts has become even more important. The contamination of spacecraft is a difficult problem that has yet to be solved because only symptomatic measures are available, such as material selection based on the amount of contaminants outgassed and baking of the materials used in spacecraft.

We have focused our research on photocatalysts to solve the contamination problem of spacecraft. When TiO_2 is irradiated with UV light, electron-hole pairs are excited and the electrons and holes diffuse to the surface, where they oxidize organic matter, ultimately decomposing it into H_2O and CO_2.

Since the report of the Honda–Fujishima effect in 1972 [53], extensive research on photocatalysts has been conducted. TiO_2 photocatalysts have already been used on roads, roofs, windows, and walls for antifouling. Although photocatalysts are commonly used on the ground, their effectiveness in the space environment, which differs substantially from the ground environment, must be confirmed.

VACUUM AND PHOTOCATALYTIC ACTIVITY

Earth orbits in which satellites fly are roughly classified into LEO and geostationary orbit (GEO). LEO is mainly used for Earth observation satellites, and the altitude ranges from 300 to 2000 km. By contrast, GEO is used for communication, broadcasting, and meteorological satellites, and the altitude extends to ~36,000 km [54]. The pressure at altitudes above 300 km is less than $~10^{-5}$ Pa.

In a vacuum environment, the absence of O_2 and H_2O affects photocatalytic activity. In the absence of H_2O, decomposition by hydroxyl radicals produced from H_2O by photocatalytic decomposition does not occur. However, the contribution of hydroxyl radicals to photocatalytic decomposition is considered to be overestimated. GuO et al. have reported that, although hydroxyl radicals contribute to decomposition, their contribution depends on the chemical structure of the pollutant [55]. They also reported that direct degradation by holes is dominant. Ishibashi et al. estimated the quantum yield for hydroxyl radicals to be 7×10^{-5}, which is smaller than the quantum yield for photocatalysis ($\sim10^{-2}$), and concluded that TiO_2 photocatalytic oxidation is mainly due to holes [56]. Therefore, the effect of H_2O is expected to be small. The role of O_2 in the photocatalytic degradation process is fourfold: (1) electron reception, (2) contribution to the degradation of active species, (3) radical chain reaction (RCR), and (4) maintenance of TiO_2 stoichiometry [57]. In the first role (electron reception), O_2 receives photo-excited electrons in TiO_2 and becomes O_2^-. The photocatalytic activity is improved because of the resultant enhancement of charge separation. In the second role (contribution to the degradation of reactive species), O_2^-, H_2O_2, and singlet oxygen ($^1O_2$) generated from O_2 are considered to participate in the degradation process. The third role, RCR, is expressed by the following equation [58-60],

$$RH + h^+ \rightarrow R\cdot + H^+,$$

$$R\cdot + O_2 \rightarrow RO_2,$$

$$RO_2 + RH \rightarrow RO_2H + R\cdot,$$

where RH is a contaminant and h^+ is the photo-excited hole in TiO_2. RO_2H is produced by oxygen and holes, and O_2 itself contributes to this reaction. In the fourth role (maintaining the stoichiometry of TiO_2), TiO_2 has been reported to gradually lose oxygen during decomposition under an O_2-deficient environment, and atmospheric O_2 has been reported to replenish the oxygen [61–63]. When O_2^- is removed from TiO_2, leaving two electrons when an O vacancy occurs, the electron density in TiO_2 increases:

$$TiO_2 \rightarrow TiO_{2-x} + \frac{x}{2}O_2 + 2x\ e^-,$$

when oxygen is added to TiO_{2-x}, the oxygen receives these electrons and becomes O_2^-, thus eliminating the oxygen vacancy.

Photocatalytic activity decreases under vacuum conditions in the absence of O_2 and H_2O. However, we considered that photocatalysts could ameliorate the contamination problem of spacecraft if a certain amount of photocatalytic decomposition occurs in vacuum. The amount of contaminants in spacecraft is extremely low compared with that in the ground environment, and the allowable contamination limit for a satellite is approximately 1 μg/cm^2. Therefore, if photocatalysts can decompose a few micrograms per square centimeter of pollutants, they can contribute to reducing the contamination problem of spacecraft.

Because of the decrease in photocatalytic activity under vacuum conditions, as previously described, few studies have investigated the use of photocatalysts in vacuum environments. To evaluate the basic photocatalytic activity under vacuum conditions, we conducted degradation experiments on methyl red (MR) dye [64]. Photocatalytic activity is typically evaluated on the basis of the decomposition of gaseous molecules or aqueous solutions; however, to mimic the mechanism of spacecraft contamination, we coated MR onto the surface of TiO_2. We found that the photocatalytic activity estimated by monitoring the decolorization of MR was comparable in air and vacuum. The effect of the vacuum environment on the photocatalytic activity should be investigated in detail, and photocatalysts with higher activity in a vacuum environment should be developed.

CONCLUSION

In this chapter, we described the space applications of TiO_2, especially the EB and AO resistance of TiO_2 and its photocatalytic activity in a vacuum environment. In addition to its ability to resist EB irradiation and AO, TiO_2 should be investigated for its ability to resist vacuum ultraviolet rays and proton beams. Elucidating the contribution of O_2 and H_2O to photocatalytic decomposition should enable the development of photocatalysts with higher activity in a vacuum environment.

REFERENCES

[1] JSME. Ed.: *JSME Mechanical Engineers' Handbook*, Applications γ11: Space Equipment and Systems, Maruzen: Japan, 2007 (in Japanese).
[2] Qin, W., Peng, D., Wub, X., Liao, J. *Mater. Chem. Phys.*, 2014 147, 311.
[3] Latthe, S. S., An, S., Jin, S., Yoon, S. S. *J. Mater. Chem. A*, 2013, 1 13567.
[4] Jun, J., Dhayal, M., Shin, J.-H., Kim, J.-C., Getoff, N. *Radiat. Phys. Chem.*, 2006, 75, 583.
[5] Kim, M. J., Kim, K.-D., Tai, W. S., Seo, H. O., Luo, Y., Kim, Y. D., Lee, B. C., Park, O. K. *Catal. Lett.*, 2010, 135, 57.
[6] Jeun, J.-P., Park, D.-W., Seo, D.-K., Kim, H.-B., Nho, Y.-C., Kang, P.-H. *Rev. Adv. Mater. Sci.*, 2011, 28, 26.
[7] Seo, H. O., Sima, C. W., Kima, K.-D., Kima, Y. D., Park, J. H., Lee, B. C., Lee, K. H., Lim, D. C. *Radiat. Phys. Chem.*, 2012, 81, 290.
[8] Kim, M. S., Jo, W. J., Lee, D., Baeck, S.-H., Shin, J. H., Lee, B. C. *Bull. Korean Chem. Soc.*, 2013, 34, 1397.
[9] Kima, H.-B., Park, D.-W., Jeun, J.-P., Oha, S.-H., Nhoa, Y.-C., Kang, P.-H. *Radiat. Phys. Chem.*, 2012, 81, 954.

[10] Pu, X., Hu, Y., Chi, S., Cheng, L., Jiao, Z. *Solid. State Sci.,* 2017, 66, 70.
[11] Bzdon, S., Goralski, J., Maniukiewicz, W., Perkowski, J., Rogowski, J., Szadkowska-Nicze, M. *Radiat. Phys. Chem.,* 2012, 81, 322.
[12] Shimosako, N., Yoshino, K., Shimazaki, K., Miyazaki, E., Sakama, H. *Thin Solid Films,* 2019, 686, 137421.
[13] Liu, N., Häublein, V., Zhou, X., Venkatesan, U., Hartmann, M., Mačković, M., Nakajima, T., Spiecker, E., Osvet, A., Frey, L., Schmuki, P. *Nano Lett.,* 2015, 15, 6815.
[14] Chen, Y., Zhao, H., Wu, Z., Huang, W., Wang, L., Guo, B. *Radiat. Phys. Chem.,* 2018, 153, 79.
[15] Di, M., He, S., Li, R., Yang, D. *Nucl. Instrum. Methods. Phys. Res. B,* 2006, 252, 212.
[16] Kimoto, Y., Miyazaki, E., Ishihara, J., Ishimura, H. *J. Vac. Soc. Jpn.,* 2009, 52, 475 (in Japanese).
[17] Verker, R., Bolker, A., Carmiel, Y., Gouzman, I., Grossman, E., Minton, T., Remaury, S. *Acta Astronaut.,* 2020, 173, 333.
[18] He, Y., Suliga, A., Brinkmeyer, A., Schenk, M., Hamerton, I. *Polym. Degrad. Stab.,* 2019, 166, 108.
[19] Kutz, M., Ed.: *Handbook of Environmental Degradation of Materials* (Third Edition); William Andrew: 2018; pp. 601.
[20] Zhang, X., Mao, L., Du, J., Wei, H. J. *J. Sol-Gel Sci. Technol.,* 2014, 69, 498.
[21] Xie, Y., Gao, Y., Qin, X., Liu, H., Yin, J. *Surf. Coat. Technol.,* 2012, 206, 4384.
[22] Chu, Y., Pan, Y., Gao, Y., Qin, X., Liu, H. *Thin Solid Films,* 2012, 526, 109.
[23] Huang, Y., Tian, X., Lv, S., Yanga, S., Fu, R., Chu, P. K., Leng, J., Li, Y. *Appl.Surf. Sci.,* 2011, 257, 9158.
[24] Minton, T. K., Wu, B., Zhang, J., Lindholm, N. F., Abdulagatov, A. I., O'Patchen, J., George, S. M., Groner, M. D. *ACS Appl. Mater. Interfaces,* 2010, 2, 2515.

[25] Ouyang, Q., Wang, W., Fu, Q., Dong, D. *Thin Solid Films*, 2017, 623, 31.
[26] Synowicki, R. A., Hale, J. S., Ianno, N. J., Woollam, J. A., Hambourger, P. D. *Surf. Coat. Technol.*, 1993, 62, 499.
[27] Qi, H., Qian, Y., Xu, J., Li, M. *Corros. Sci.*, 2017, 127, 56.
[28] Xiao, F., Wang, K., Zhan, M. *J. Am. Ceram. Soc.*, 2012, 123, 143.
[29] Liu, B., Pei, X., Wang, Q., Sun, X., Wang, T. *Surf. Interface Anal.*, 2012, 44, 372.
[30] Duo, S., Li, M., Zhu, M., Zhou, Y. *Surf. Coat. Technol.*, 2006, 200, 6671.
[31] Shimamura, H., Nakamura, T. *Polym. Degrad. Stab.*, 2009, 94, 1389.
[32] Miyazaki, E., Tagawa, M., Yokota, K., Yokota, R., Kimoto, Y., Ishizawa, J. *Acta Astronaut.*, 2010, 66, 922.
[33] Yokota, K., Abe, S., Tagawa, M., Iwata, M., Miyazaki, E., Ishizawa, J.-I., Kimoto. Y., Yokota, R. *High Perform. Polym.*, 2010, 22, 237.
[34] Gouzman, I., Girshevitz, O., Grossman, E., Eliaz, N., Sukenik, C. N. *ACS Appl. Mater. Interfaces*, 2010, 2, 1835.
[35] Tsai, M.-H., Liu, S.-J., Chiang, P.-C. *Thin Solid Films*, 2006, 515, 1126.
[36] Shimosako, N., Hara, Y., Shimazaki, K., Miyazaki, E., Sakama, H. *Acta Astronaut.*, 2018, 146, 1.
[37] Barnes, W. L., Xiong, X., Salomonson, V. V. *Adv. Space Res.*, 2000, 32, 2099.
[38] Teillet. P. M., Slater, P. N., Ding, Y., Santer, R. P., Jackson, R. D., Moran, M. S. *Remote Sens. Environ.*, 1990, 31, 105.
[39] Wright, R., Rothery, D. A., Blake, S., Harris, A. J. L., Pieri, D. C. *Geophys. Res. Lett.*, 1999, 26, 1773.
[40] Flynn, L. P., Harris, A. J. L., Wright, R. *Remote Sens. Environ.*, 2001, 78, 180.
[41] Bessho, K., Date, K., Hayashi, M., Ikeda, A., Imai, T., Inoue, H., Kumagai, Y., Miyakawa, T., Murata, H., Ohno, T., Okuyama, A., Oyama, R., Sasaki, Y., Shimazu, Y., Shimoji, K., Sumida, Y.,

Suzuki, M., Tanihuchi, H., Tsuchiyama, H., Uesawa, D., Yokota, H., Yoshida, R. *J. Meteorol. Soc. Japan,* 2016, 94, 151.

[42] Shiomi, K., Kawakami, S., Kina, T., Yoshida, M., Sekio, N., Mitomi, Y., Kataoka, F. *Proc. Envisat Symposium,* 2007, ESA SP-636.

[43] Takahashi, T., Mitsuda, K., Kelley, R. *AIP Conf. Proc.,* 2010, 1248, 537.

[44] Nakagawa, T., Murakami, H. *Butsuri,* 2004, 59, 3 (in Japanese).

[45] Hikida, R., Yoshioka, K., Murakami, G., Kuwahara, M., Yoshikawa, I. *JAXA Research and Development Report,* 2002, JAXA-RR-16-012 (in Japanese).

[46] Kurihara, S., Murakami, H., Tanaka, K., Hashimoto, T., Asanuma, I., Inoue, J. *Proc. SPIE,* 2002, 5234, 11.

[47] Urayama, F., Yano, K., Yamanaka, R., Miyazaki, E., Kimoto, Y. *Journal of the Japan Society for Aeronautical and Space Sciences,* 2008, 56, 543 (in Japanese).

[48] Bando, T., Hara, H., Urayama, Kimoto, Y., Miyazaki, E. *Proc. 55th Space Sciences and Technology Conference,* 2011, JSASS-2011-4254 (in Japanese).

[49] Darrin, A., O'Leary, B. L. *Handbook of Space Engineering, Archaeology, and Heritage,* CRC Press, 2009.

[50] Bonnet, R. M., Lemaire, P., Vial, J. C., Artzner, G., Gouttebroze, P., Jouchoux, A., Leibacher, J. W., Skumanich, A., Vidal-Madjar, A. *Astrophys. J.,* 1978, 221, 1032.

[51] Jiao, Z., Jiang, L., Sun, J., Huang, J., Zhu, Y. *IOP Conf. Ser.: Mater. Sci. Eng.,* 2019, 611, 012071.

[52] Dai, W., Qiu, J., Shen, Z., Yang, Y. *Adv. Astronaut. Sci. Technol.,* 2018, 1, 183.

[53] Fujishima, A., Honda, K. *Nature,* 1972, 238, 37.

[54] Ohnishi, A. et al. Ed: *Thermal Design of Spacecraft.* The University of Nagoya Press: 2014 (in Japanese).

[55] Guo, M. Y., Ng, A. M. C., Liu, F., Djurišić, A. B., Chan, W. K. *Appl. Catal. B,* 2011, 107, 150.

[56] Ishibashi, K., Fujishima, A., Watanabe, T., Hashimoto, K. *J. Photochem. Photobiol. A,* 2000, 134, 139.
[57] Fujishima, A., Zhang, X., Tryk, D. A. *Surf. Sci. Rep.,* 2008, 63, 515.
[58] Ohtani, B., Nohara, Y., Abe, R. *Electrochemistry,* 2008, 76, 147.
[59] Heller, A. *Acc. Chem. Res.,* 1995, 28, 503.
[60] Schwitzgebel, J., Ekerdt, J. G., Gerischer, H., Heller, A. *J. Phys. Chem.,* 1995, 99, 5633.
[61] Yoshida, K., Nanbara, T., Yamasaki, J., Tanaka, N. *J. Appl. Phys.,* 2006, 99, 084908.
[62] Muggli, D. S., Falconer, J. L. *J. Catal.,* 1999 187, 230.
[63] Muggli, D. S., Falconer, J. L. *J. Catal.,* 2000, 191, 318.
[64] Shimosako, N., Sakama, H. *Acta Astronaut.,* 2021, 178, 693.

In: Titanium Dioxide
Editor: Aparna B. Gunjal
ISBN: 978-1-68507-457-9
© 2022 Nova Science Publishers, Inc.

Chapter 2

REVOLUTION OF TITANIUM DIOXIDE IN BIOMEDICAL AND APPLICATIONS IN ENVIRONMENTAL REMEDIATION

Smita Kumari[1,*] *and Dharmendra Kumar*[2]
[1]Environmental Biotechnology Division, CSIR-Indian Institute of Toxicology Research, Lucknow, Uttar Pradesh, India
[2]Department of Electronics and Communication Engineering, Madan Mohan Malaviya University of Technology, Gorakhpur, India

ABSTRACT

Multidisciplinary research has led a tremendous achievement to mitigate the complications of various fields such as in medical, environment, energy production, and biosensing. One such effective technology is titanium dioxide (TiO_2), known to be a great possible candidate in various aspects of biomedical fields and environmental remediation purposes deprived of harmful effects. The nanostructures of TiO_2, nanowires, and nanotubes, have successfully been employed owing to astounding properties such as excellent biocompatibility, large

---

[*] Corresponding Author's E-mail: smitakumari14@gmail.com.

surface vicinity, non-toxicity, lightweight, ceramic, polymeric, strong photocatalytic ability, stability, and slightly cheaper than other semiconductor materials. However, the scarcity of water has emphasized the application of nanostructured TiO_2 as a better option to accomplish the urgent necessity of pure drinking water due to its remarkable photocatalytic function. In this chapter, we have discussed the improvement of environmental concerns and how TiO_2 technology is effectively employed for the degradation of organic pollutants such as pesticides, and antibiotics for wastewater treatment. Furthermore, we have elaborated the application and role of TiO_2 indiscriminately in bones implantation, antibacterial activity, and immobilization of various drug carriers. In addition, we have highlighted TiO_2 photocatalyst in photodynamic therapy, is a promising approach for cancer therapy.

Keywords: titanium dioxide, photocatalyst, organic pollutants, environmental degradation, antibacterial activity, photodynamic therapy, implantations

INTRODUCTION

The emerging field of nanotechnology in various applications has insight into a great opportunity in multidisciplinary research. Nanotechnology gains visibility among the researchers to create the materials in nano form. To date use of numerous materials in the form of nanomaterials have proven its potential in the wide area of biomedical in pharmaceutical products and medicines, agriculture, biosensor, environmental applications like degradation of chemical pollutants for water purification, and so on. Out of those, titanium dioxide (TiO_2) is one of the most demanding and applied technologies in biomedical as well as in environmental remediation applications. Titanium was first discovered by William Gregor in the year 1791. The excellent properties of TiO_2 make it an effective and promising technology. In comparison to other nanomaterials used, TiO_2 is comparatively cheap, nano-sized, biocompatible, low toxicity, resistant to corrosion, and has high chemical stability (Akakuru et al. 2020; Das et al. 2021). The photocatalytic property of TiO_2 is a very significant characteristic because of its applications in both medical and environmental. TiO_2 on ultraviolet (UV) irradiation generated various reactive oxygen species (ROS). These highly reactive species oxidized environmental contaminants such as pesticides, heavy metals, and antibiotics, and damage biomacromolecules like DNA.

Nowadays, this technology is widely applied at an industrial scale due to its crystalline structures. The TiO_2 consists of two different crystalline types i.e., Anatase and Rutile are extensively used in biomedical owing to their cytotoxic and photocatalytic properties. ROS induced cytotoxicity on damaging cell membrane via TiO_2 photocatalysis. Moreover, the photocatalytic property of TiO_2 has potential application in photodynamic therapy (PDT) in which some nonmalignant conditions and cancer treatments are performed (Çeşmeli and Biray Avci, 2019). The capability of TiO_2 is enhanced when it functions as nanocomposites and act as a potential device for drug delivery, cell imaging, genetic engineering, and biological analysis. Further, TiO_2 is a semiconductor that functioned as a biosensor in diseases diagnosis is discussed later (Mavrič et al. 2018).

However, in recent decades, TiO_2 technology has had huge applications in the field of environmental research for the degradation of organic pollutants in wastewater such as dyes, pesticides, and antibiotics. Recently, photocatalysis has gained great popularity in the field of environmental remediation (Wen et al. 2015). The various semiconductors have been used for the treatment of wastewater and air purification by employing photocatalysis. Owing to the high stability, large surface area, and strong photocatalytic ability, TiO_2 is proven one of the effective semiconductors for wastewater treatments in comparison to other conventional and membrane-based technologies. Besides, direct use of TiO_2 photocatalyst in an aqueous medium can be immobilized in a reactor system of photocatalytic to prevent the turbidity occurring in the aqueous medium due to catalyst presence which led to a decrease in diffusion of UV light in the solution. TiO_2 has wide application in various areas that we will discuss further (Figure 1).

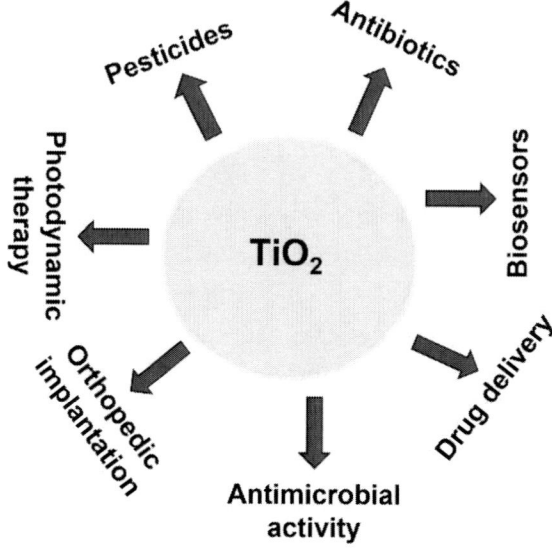

Figure 1. Overview on the application of TiO$_2$ in biomedical and environmental remediation.

ROLE OF TIO$_2$ FOR ENVIRONMENTAL REMEDIATION

The rapid increase in urbanization and industrial development led to an increase in the scarcity of water due to organic contamination of water resources. On the other hand, due to the expansion of the human population, the water demand also increases. To overcome this scarcity various techniques have emerged but with little effect. In recent days, nanomaterials are largely used in research for the decontamination of environmental pollutants. However, in this context, TiO$_2$ nanomaterial exhibits a major role in the degradation of organic contaminants owing to its remarkable properties of photocatalysis activity. Hence, further, we will discuss the role of TiO$_2$ in wastewater treatment on the mitigation of pesticides and antibiotics. We also emphasize the mechanism of the TiO$_2$ photocatalyst and its antibacterial activities.

Mechanism of TiO₂ as a Photocatalyst

We have mentioned here the valuable properties of TiO_2 which is most significant for photocatalytic application owing to low cost, easy immobilization on various surfaces, high oxidizing capacities, and nontoxicity. In general, the bandgap of TiO_2 semiconductors ranges from 3.0 to 3.2 eV along with high absorption capacity in the UV region and can be tunable for reducing bandgap energy. The reduced bandgap increases the efficiency of TiO_2 as a photocatalyst (Zahoor et al. 2018). In addition, TiO_2 on ultraviolet (UV) irradiation generated various reactive oxygen species during photocatalytic activity. For instance, hydrogen peroxide (H_2O_2), hydroxyl radicals (OH·), superoxide ($O_2^{-\bullet}$), and singlet oxygen ($^1O_2$) (Varma et al. 2020). The basic mechanism of photocatalysis revealed that when TiO_2 semiconductor is illuminated in UV light then it makes a hole (h^+) in the valence band and excited the electron (e^+) in the conduction band to produce ROS which is responsible for the oxidation of organic contaminants into the simplest form (Figure 2). Thereafter, we will discuss the applications of TiO_2 in the removal of pesticides in drinking water, and groundwater.

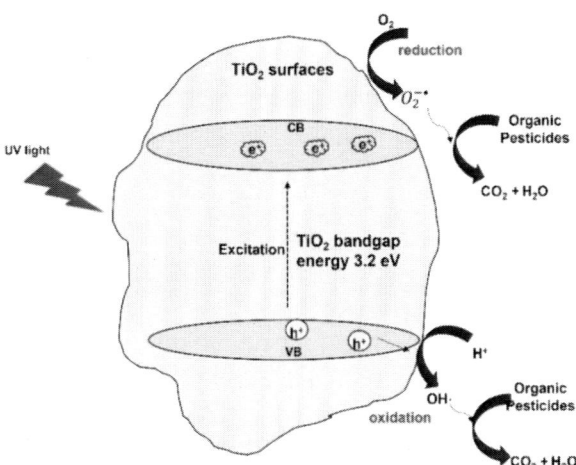

Figure 2. Schematic diagram showing the basic principle of TiO_2 photocatalytic degradation of contaminants or pesticides into simplest form.

Advances of TiO$_2$ Technology in Pesticides Degradation

In this chapter, we will discuss the degradation of various pesticides that are released through anthropogenic activities such as industrial effluents and agriculture runoff. These released pesticides merge into the various water source and endanger the life of the aquatic ecosystem as well as human health. Pesticides that are persistent are known as persistent organic pollutants (POPs) for example, polychlorinated biphenyls (PCBs). Such POPs persist in the environment for a long time and cannot degrade readily. These toxic pesticides enter the human organism through various routes. Through skin contact, ingestion, drinking water, and consuming food and breathing they are carried via air on vaporization and as dust particles through water sources and persist longer in the human organs. Most pesticides are endocrine disruptors, carcinogenic and reproductive defects in human and aquatic organisms like fish. Various conventional methods have been applied for the removal of POPs for water treatment but failed to achieve the complete enhanced removal rate. Advance methods using nanomaterials have been carried out by using different materials but among them, TiO$_2$ is proven to be a better technology for wastewater treatment. TiO$_2$ is prepared by several methods, such as the sol-gel method, which is not discussed in this chapter, which has high photocatalytic activity results from effective degradation of pesticides applied for agriculture in many ways. Few pesticides, for instance, 2,4-di-chlorophenoxypropionic acid (2,4-DP) and 2,4-di-chlorophenoxyacetic acid (2,4-D) are used in cereals agriculture, Carbamate pesticides for pest control, triazines like Atrazine, phosphorus-based pesticides, and N-based pesticides are effectively degraded by a photocatalyst, TiO$_2$. Some common pesticides used on fruits and vegetables crops are Fludioxonil, Azoxystrobin, Cyproconazole, Metalaxil, Pendimethalin, Metribuzin, Propyzamide, and Rimsulfuron are degraded by TiO$_2$ as an aqueous suspension with Na$_2$S$_2$O$_8$ in the presence of sunlight into the non-toxic form including water, inorganic ions, and carbon dioxide (Kanan et al. 2020).

Using bare TiO$_2$ creates some post-treatment problems like recovery of the photocatalyst. To overcome this issue TiO$_2$ has been explored to be immobilized with the different solid systems, like SBA-15 silica gel, which not only solved the post-treatment recovery but can also ease the renewable use of photocatalyst (Liou et al. 2018). Generally, pesticides degradation routes follow Langmuir–Hinshelwood kinetic model where the products are non-toxic (Amalraj and Pius, 2015). Pesticide degradations rely on several factors mostly catalyst concentration, pH, electron acceptor, and substrate concentrations. The degradation efficacy of TiO$_2$ as pristine is proven potential but in combination or addition of nanomaterials (graphene) as nanocomposites, surfactants (cetyltrimethylammonium bromide), oxidants like peroxydisulfate, and inorganic compounds (Na$_2$S$_2$O$_8$) enhanced the efficiency of catalyst multiple times than the pristine photocatalyst (Luna-Sanguino et al. 2018; Abdennouri et al. 2016). Therefore, TiO$_2$ technology plays a promising role in the degradation of organic contaminants with high efficiency.

TiO$_2$ Mitigates Antibiotics for Wastewater Treatment

Medicines or drugs especially antibiotics becoming life supporters for the health and wellbeing of society. The development of the different types of medication reduced the mortality rate and improves the expectancy of life. However, large quantities of antibiotics enter the environmental water bodies, like groundwater, surface water, drinking water, and seawater during the fermentation and purification process while production and consumption. Industrial effluents contain tons of residual antibiotics as wastewater. The most worrisome thing is that these antibiotics alter the genetics of the bacteria present in a bioreactor which produces bacteria resistant to multiple antibiotics. When these wastewaters discharge with anti-microbial resistance in the aquatic system it poses a significant threat when consumed as drinking water by human beings. Consequently, this largely affects the water sources and results threat to the environment. Antibiotics are considered a micropollutant, and the individual antibiotic remains in the water as nano to microgram per liter with minimal effect, but in the natural environment, it exists in the mixture and might have the greater combined effect (Kovalakova et al. 2020). Antibiotics in the environment are undesirable due to being non-biodegradable, toxic, highly persistent, and bioaccumulative (Kovalakova et al. 2020; Prashanth et al. 2021). Although many conventional treatment processes have been applied, they fail to be degrading persistent antibiotics. Therefore, to eradicate the toxic antibiotics from the water sources or to treat wastewater, advanced technology such as membrane filtration, adsorption technology, and advanced oxidation processes are applied. On the other side, it must verify that the technologies should be effective, and the final products in the process must be non-toxic. But there is one considerable drawback in the membrane filtration and adsorption technology that the final product still preserves the toxicity which is again another issue for its removal. Therefore, advanced oxidation processes (AOPs) are considered excellent and effective technologies due to the production of simple and non-toxic final products. The most common AOPs are electrochemical oxidation, Fenton process, ozonation, and heterogeneous photocatalysis (Giwa et al. 2021). All these processes are common in one respect that they all produced highly reactive species, a game-changer species in the removal of antibiotics for wastewater treatment. Among the above technologies, there is extensive use of heterogeneous photocatalysis owing to sustainability. Several semiconductor materials perform heterogeneous photocatalysis and TiO_2 is one of them because of the unique properties (as discussed earlier) that serve as a potential heterogeneous photocatalyst (Varma et al. 2020).

Heterogeneous Photocatalysis by TiO_2

However, it is very important to make sustainable and cost-effective processes if applied at a commercial scale. TiO_2 has a wide bandgap and so that can only harvest UV light effectively and this limits the photocatalytic activity of TiO_2. For sustainable photocatalytic function, various surface modifications are performed so that the bandgap of TiO_2

is reduced, and it can be able to absorb visible light instead of UV only. Doping of TiO_2 with metal, non-metal like S, F, and coupling with semiconductors such as ZnS, CdS, etc can overcome these limitations (Wetchakun et al. 2019; Prashanth et al. 2021; Kumar et al. 2020). Various factors that influence the efficacy of photocatalytic degradation of antibiotics are the effect of antibiotic concentration, catalyst concentration, pH, other inorganic and minerals present, and intensity of light sources. Antibiotics that frequently contaminate water sources are spiramycin, sulfamethoxazole, trimethoprim, and ciprofloxacin (Kutuzova et al. 2021). Tiwari et al. (2018) studied the degradation of sulfamethoxazole on nanocomposite film Ag0(NP)/TiO_2. They observed that the degradation efficiency decreases by increasing the concentration of antibiotics. Moreover, on increasing the concentration of photocatalyst the degradation rate is simultaneously increased due to enhanced production of reactive oxygen species. For the pre-treatment of wastewater contaminated with spiramycin, $TiO2/SO4$ solid superacids can be a promising method for breakdown at the molecular level (Yang et al. 2019). TiO_2 in combination with biochar is reported on the same conclusion for the degradation of sulfamethoxazole under heterogeneous photocatalytic degradation and proved better than homogenous photocatalytic degradation (Kim and Kan, 2016).

Purpose of Surface Characterization of TiO_2

In the last few eras, there are rapid evolutions of nanomaterials in various fields. The basic steps to insight the chemical and physical phenomena implicated at nano and micro level surface characterization. Crystal of TiO_2 can be studied in three different phases, rutile, anatase, and brookite and each phase is unique on its own in terms of electronic and optical properties (Yasutake et al. 2021). The surface energy of TiO_2 relies on size and composition which define their contact with pollutants or other solid substrates. The size of the material can considerably affect the photocatalytic property of TiO_2 (Li et al. 2015) which is the key property, consequently, modifying the oxidation of pollutants such as pesticides and antibiotics in water bodies. Extensive use of TiO_2 at micro and nanoscale in industries and for commercial purposes in comparison to larger size confers a high surface area that exhibits high reactivity (Nam et al. 2019). The above characteristics have great importance in biomedical applications too.

APPLICATIONS OF TIO_2 IN BIOMEDICAL

As of now, we know that functional material, TiO_2 is extensively used in environmental pollutant remediation and antibacterial activities. Apart from these, TiO_2 material alone or in nanocomposites form is used as a promising technology in the biomedical application which performs a significant role in improving health such as in bone implantation, dental care, and cancer treatment. This also acts as a vector for drug delivery and released the drugs at the target site. The unique biological and physicochemical properties of TiO_2 material resulted in its use in various health care areas. Besides, the unique properties of TiO_2 as discussed earlier, and excellent photocatalytic characteristic makes it an important and efficient candidate for various biomedical applications that we would discuss next.

SUSTAINABLE DRUG RELEASE STRATEGIES

The conventional methods of drugs delivery processing are normally by oral, inhalation routes, and parenteral where drugs are dispersed to the entire body and not to the specific site of concern. The nonspecific drugs administration makes the method limited that's why the call for the establishment of advanced technology in biomedical is important. In this view, robust interdisciplinary strategies with combined efforts of material engineers, biologists, and medical scientists have recently proven promising findings. There is extensive use of nanocomposite materials or nanomaterials for drug delivery and biomedical applications. Due to their distinctive properties, a lot of research activities have been focused on different morphologies of TiO_2 such as nanotubes or nanopores materials from transition metal oxides, and with pompon-like porous TiO_2.

Titanium Dioxide Nanotubes (TNTs)

Titanium nanotubes (TNTs), because of their outstanding biocompatibility, are extensively applied in drug delivery systems. Fabrication of TNTs with suitable nanotube structures is performed by an electrochemical anodization process. The simple electrochemical arrangement consists of two electrodes one is containing titanium and the other being counter electrodes that are submerged in electrolytes connected by an external current. In the initial process of electrochemical anodization, metal dissolution and formation of oxide take place on the surface of the titanium. Further, TiO_2 layers with self-ordered and vertically aligned tubular oxide or porous are formed. However, by altering the electrochemical anodization factors such as time, anodizing voltage, and compositions of the electrolytes, TNTs dimensions could be controlled (Manivasagam et al. 2021). To diminish the adverse effects on the healthy tissues and to boost up the therapeutic effects there must be a safe and accurate strategy to deliver the drugs at the target sites during different types of treatment therapies. For this, TNTs efficiently stimulate the development of drug-releasing for the treatment of cancer, bone-related diseases, and post-implant infections. TNTs structures are a flexible and perfect platform for the drug-releasing strategies which depends upon on-demand release, short to long release, or time-programmed release with specific or multiple drug payloads. Because TNTs are able to carry different types of drugs and release them in different conditions for different therapies in specific body parts, they are proved better over the conventional drug delivery techniques (Wang et al. 2017). Furthermore, the performance of TNTs could be enhanced by investigating different strategies. Recently, nano-drug delivery systems for various implantations and cancer therapies create a challenge owing to low bioavailability, their inadequate cellular uptake, weak targetability, and stability concerns. Stimuli-responsive systems such as pH-based strategies and enzyme-based strategies are applied nowadays to deliver a drug locally.

pH Stimuli Dependent Strategies

To function the TNTs in a controlled manner, the strategy opted to explore it in different pH-responsive polymers because drugs released through pristine titanium are a very fast process. Therefore, considering pH, pure TiO_2 and polymer/TiO_2 nanotubes composites have been examined for best-fit parameters for a sustainable drug delivery approach. For nanocomposites study, polyethylene glycol (PEG) and poly (lactic-co-glycolic acid) (PLGA) polymer have been used in combination with TiO_2 (Du et al. 2015). Carprofen and lidocaine drugs are examined for delivery through these strategies, but both drugs can act differently at different pH levels and solubility of drugs. The mechanism of drug delivery through PLGA/TiO_2 nanocomposites is not solely dependent on the solubility of drugs at different pH but can also be liable on two-stage mathematical models. The two models that are used for best-fit parameters i.e., first-order and Gallagher–Corrigan (Cui et al. 2021). The first one is the first-order kinetic model which is explained in equation 1, where cumulative drug release amount is Mt/M∞ at time t and infinite time, k is the first-order release constant, b is constant:

$$M_t/M_\infty = 100 - e^{b-kt} \tag{1}$$

The first-order kinetics is for dispersion-controlled dissolution of the drug into the medium.

The second mathematical model is Gallagher–Corrigan as in equation 2:

$$f_t = f_B \cdot \left(1 - e^{-k_1 t}\right) + (f_{tmax} - f_B) \cdot \left(\frac{e^{k_2 t - k_2 \cdot t_{2max}}}{1 + e^{k_2 t - k_2 \cdot t_{2max}}}\right) \quad (2)$$

where f_t; accumulative drug release fraction at time t, k_1; the first-order release constant (Stage 1), k_2; the second stage release constant owing to the polymer degradation, f_B; accumulative drug release % during the Stage 1, f_{tmax}; maximum drug release % during the whole process, t_{2max}; time at which drug release rate reaches the maximum. This model state that the release of drugs solely depends on the degradation of polymer used.

Low pH is effective in releasing antibacterial agents via TNTs on the infectious site during osteogenic therapies. One of the studies by Wang et al. (2017) reported that a hybrid system, metallic ion coordination polymer with TNTs, in which antibiotics agents can be inserted into TNTs and sealed by coordination polymers. This system is applied at the infectious site caused by *Staphylococcus aureus* and *Escherichia coli* which form an acidic environment at the site. The used metallic ions in the hybrid system, Zn^{2+} or Ag^+, form a strong coordination bond between coordination polymer and TNTs. At the onset of infections, the environment becomes acidic and consequently, breaks the coordination bond because the metallic ions are sensitive to low pH and enhanced release rate of antibacterial agents. The surface of TNTs is modified with Fe^{3+} and its ease to bind orthopedic treatment drug alendronate sodium up to 50% by weight. Upon the pH shift the Fe^{3+} break and release the drug from the surface of TNT in a controlled manner for a prolonged period (Liu et al. 2019).

Enzyme Stimulates Drug Delivery

However, pH-triggered drug delivery strategies have been widely applied on Ti implants in various medical treatments but there are certain mortality and failure associated with implant-related infection due to the biofilm formation (Cloutier et al. 2015). Different approaches are still under the experiments which are expected to mitigate such issues. Therefore, in this context, we will discuss how the bacterial enzyme stimulates in releasing the antibiotics drugs loaded on Ti. An enzyme secreted from *S. aureus* is a good example because it causes severe infection at the material implant surface (De Breij et al. 2016). The surface of the titanium implant is modified by attaching with peptide conjugated vancomycin drug. Once the bacteria infected the cell it secreted specific serine protease-like protease (SplB) enzyme which stimulate the drug to release by cleaving the attached peptide from the titanium. Peptides are sensitive to the specific enzyme, and it only cleaves when the infection occurs, and the drug works. The increased concentration of the enzyme decided the effect of antibacterial (Zhang et al. 2021). Chitosan and Hyaluronidase (HAase) are other reported enzymes released upon the onset of the bacterial infection at the implant surface. The multilayer coating of polymer conjugates with drug deferoxamine (DFO) packed TNTs is hyaluronic acid grafted gentamicin (HA-gen). In the presence of enzyme HAase, HA-gen gets cleaved, reduced the microbial growth at the implant surface, and facilitates the release of 70% DFO within 72 hours resultant in the acceleration of osteogenesis and angiogenesis (Yu et al. 2020). Enzyme triggered system achieved great advantages in the context of bacterial infection occurring at implant surface which facilitates the burst release of drugs in a short time, no external stimulus required which is a big achievement in organ implant via materials in medical sciences.

Porous TiO$_2$ Microspheres in Drug Delivery

Here, we will discuss one more different morphology of TiO$_2$ which shows great potential in sustainable drug release strategies. Porous TiO$_2$ morphology has pompon-like spheres which influence drug loading pattern and its sustainable release. This porous sphere is ~2.0 μm in diameter, has a large pore volume, large surface area, and has titanium and oxygen elements as chemical composition. To understand the drug-releasing pattern via porous TiO_2 microsphere, doxorubicin hydrochloride (DOX) is used as a model drug. Molecules of DOX get trapped inside the pores and released in a controlled way via the diffusion process. DOX molecules form hydrogen bonds on the porous TiO_2 microsphere. The drug release through the porous TiO_2 microsphere could be via two different stages, one is a fast release and the second through the continuous release stage. Time taken in the former stage is 3 hours whereas, the later stage lasts for 21 hours to release. Both stages are equivalently important in the disease like cancer therapies and similar life-threatening diseases. Because initially, these kinds of severe problems need fast requirement of drugs in large scale at the target site which reduces the abnormal growth of cell and later the continuous effect of drugs progressively kills the cancerous like cells. It shows an excellent sustainable drug release property. The drug release strategy of porous TiO_2 pompon-like spheres is like that of TNT, but TNT shows better than this because the drugs release period is longer. This strategy also shows that the drug release fitted for the first-order kinetic equation (Cui et al. 2021).

APPLICATIONS OF TiO_2 AS A BIOSENSOR

The term biosensor was initially introduced by the scientist Leland Clark in 1962 and he is known as 'Father of Biosensor'. A biosensor in which the intensity of signal generation is completely reliant on the interaction of biorecognition element with the analyte and its concentration is generally known as an electrical biosensor. Electrical biosensors are divided into two classes for biosensing applications; one is an electrochemical sensor and the other is the electronic sensor. But due to the extensive use of reagents, high cost, longer time is taken, and complex processes in electrochemical, researchers began to investigate electronic approaches. Electronic biosensors are designed as electronic devices basically on micro to nano-sized to detect the chemical substances and transmit the detected substances into an electrical signal (Vu and Chen, 2019). The nanosized hybrid material facilitates great impact in developing biosensors as it increased output signals, retaining prolonged biomolecular function. Widely used conducting materials TiO_2 for biosensing application is a good example owing to its high electrical stability, potential in electron mobility, easy availability, biocompatibility, ecofriendly, and cost-effective makes it an effective candidate for biosensing application (Mavrič et al. 2018; Wang et al. 2015).

Nano-based TiO_2 enhances the performance of biosensors as it provides the accessibility of surface modification and easily immobilized biomolecules such as enzymes which enhance the sensitivity of biosensors (Shetti et al. 2019). It has high efficiency to attach with several bio-recognition components. TiO_2 integrates with several other nanoparticles for nano-hybrid biosensor to strengthen the binding with bio-recognition components. Few TiO_2 based sensors are highlighted. The enzymatic glucose sensor is developed using TiO_2/ carbon nanotubes-CO_3O_4 is sensitive to visible light and can be immobilized with glucose oxidase (Çakıroğlu and Özacar, 2018). Another hybridized material is poly-3-hexylthio-phene (PHT) developed as TiO_2/PHT. This nanohybrid also acts as an electrical biosensor (Kadian et al. 2018). Despite this, TiO_2 is used as a sensor for breast cancer in combination with reduced graphene oxide (Safavipour et al. 2020). Due to the development of such sensitive sensors, the diagnostic and therapy of critical diseases become easier.

TiO_2 IN PHOTODYNAMIC THERAPY

Diseases like cancer have threatened communities, yet there is not a single therapy that exists and eradicates cancer. Still, to get relief to some extent, many technologies have been experimented with in biomedical fields. Among those technologies, nanomaterials are applied at a high level and receive good consequences. Nanomaterial TiO_2 is gaining much attention among medical researchers because of its effective and favorable properties especially, photocatalytic (discussed earlier). Various classical therapies have been applied for cancer treatment, for instance, radiotherapy, surgery, and chemotherapy but during the treatment when these therapies are applied healthy cells get damaged causing severe cytotoxicity, and chances of recurrence of the problem are high (Hur et al. 2017; Behranvand et al. 2021). Moreover, photodynamic therapy (PDT) is an advanced method that uses photosensors and light in cancer therapy without any negative biological impact on its own (Kwatra and Mudgil. 2020). TiO_2 is synthesized in nanoscale with modification on the surface is used as an active photosensitizer agent for photosensitization in PDT for breast cancer, kidney, and cervical cancer (Yurt et al. 2018; Yang et al. 2021). It functions only on one principle that when UV is illuminated on the photosensitizer it activates and generates reactive oxygen species because of photocatalytic properties and kills the cancerous cells. PDT is not merely confined to cancer treatment, but it is also applied in various other diseases such as diabetes mellitus, rheumatoid arthritis, antimicrobial therapy, and muscular degeneration (Kou et al. 2017).

The toxicity of pristine TiO2 is not clear except when it is used as pristine it generates ROS on its own after irradiation, but its function is significantly increased when it is combined with antibodies, enzymes, nanomaterials, or other compounds such as derivatives of porphyrin. Porphyrin is considered a photosensitizer and toxic when it is introduced alone for therapy but significantly enhances the potential of the TiO_2 for medical application by reducing self-toxicity. Water-soluble 5,10,15,20-tetra(4-sulfonatophenyl) porphyrin (TSPP) in combination with TiO_2 as a suspension is used for the treatment of rheumatoid arthritis in mice and rats (Zhao et al. 2015). During the PDT, nanocomposite has significantly affected by declining the INF-α and IL-17 in the blood serum, the key cytokines for the cause of the autoimmune disease. The fluorescence image obtained from the device under a specific wavelength of 520 nm revealed that the intense fluorescence was only achieved from sick joints.

PDT has experimented with the treatment of type II diabetes mellitus in the mouse. TSPP-TiO$_2$ suspension is injected into a mouse model and irradiated under visible light (500 to 550 nm) for 1 hour. Two hours after exposure it is observed that the blood sugar level is diminished up to 33% because of the generation of ROS (Rehman et al. 2016). Hence this combination can be effective in the treatment of type II diabetes mellitus.

TiO$_2$ in Orthopedic Implantations

During the implantation of solid tissues, various factors are required to accomplish for normal physiological behavior of the human body. Recently, TiO$_2$ nanomaterial brings a revolution in surgical orthopedic implantations. Because of good biocompatibility, high strength, good corrosion resistance, stability, and low elastic modulus, TiO$_2$ represents a promising candidate for various implantations in the artificial knee joint, femoral joint, elbow joint, etc. However, to make nanomaterial functional in a physiological environment numerous physical, chemical, or surface modifications are needed. The modifications enhance the osteointegration, osteogenesis, osteoblasts adhesion, and increase antibacterial properties (Lai et al. 2017; Raines et al. 2019). TiO$_2$ implantation can show an effective result only after surface modifications because maximum biological reactions occur between implants and tissues. To achieve improved osteointegration and osteoblast, cell proliferation titanium surface can be modified by creating a porous structure. Nanotube structure is formed upon the Physico-chemical modification on the titanium surface by the technique known as anodization (Gong et al. 2019). Different other methods for surface modification of TiO$_2$ are plasma treatment (Chiang et al. 2018), hydrothermal treatment (Vishnu et al. 2020; Yang et al. 2020), and bioactive surface (Yang et al. 2017).

Anodization is a rapid process to make TiO$_2$ nanotubes which are used on Ti-based devices. In this technique, an electrochemical treatment occurs in an electrolytic solution where the substrate is treated in the anodic chamber of the cell. During the process of anodization, an electrochemical reaction happens, and the dimensions of the nanotubes are controlled by altering a few parameters (as discussed in TNT). The TiO_2 nanotube can encourage bone tissue growth and hence stabilize the bonding between bone tissue and implant (Awad et al. 2017; Khoshroo et al. 2017). However, osteogenesis and antibacterial are two important requirements for orthopedic implants. To elevate the antibacterial property of the TiO_2 nanotube several metallic nanoparticles like strontium, silver, and gold are incorporated on its surface to fight against post-operative bacterial infection (Yang et al. 2016; Viet et al. 2018; Pan et al. 2020). Hydrophilicity i.e., water contact angle, of biomaterial, is also important to assess the biocompatibility because at the early stage of the treatment hydrophilic surface has a propensity to encourage cell proliferation and adhesion (Blatt et al. 2018).

Besides this, TiO_2 nanotubes and nanofilms have broad application in enhancing hemocompatibility and various medical devices such as prosthetic heart valves, cardiovascular stents, etc. (Junkar et al. 2020; Jiang et al. 2020). TiO_2 nanotubes are highly effective in dental implantation and enhance its property by incorporating silver and zinc nanoparticles on the surface of a titanium implant (Roguska et al. 2018).

Antimicrobial Activity of TiO_2

The increasing risk of bacterial growth in the food sector, biomedical, and water has raised great concern for health and the environment. Besides the application of pristine TiO_2 in pesticides and antibiotics removal from the environment, it also exhibits antibacterial activity. The antibacterial activity of TiO_2 is due to its nano size, concentration, stability, and surface chemistry which significantly influence the photocatalytic property and retention time in the interaction of bacteria and nanomaterials (Barreca et al. 2010; Sánchez-López et al. 2020). The antibacterial activity of TiO_2 works on the basic concept of photocatalysis. The production of ROS under UV light could initiate the damage of bacterial cell parts, like nucleic acid, and protein (Zhao et al. 2020). TiO_2 nanomaterial and nanotubes, owning sterilization features in presence of UV light, restrain the undesirable growth of bacteria such as *Escherichia coli*, *Proteus vulgaris*, *Pseudomonas aeruginosa*, and *Staphylococcus aureus* in water, food products, and from the environment (Podporska-Carroll et al. 2015; Albukhaty et al. 2020; Mahboob et al. 2021). The combination of TiO_2 with 5, 10, 15, 20-tetrakis (2,6-difluoro-3- sulfophenyl) porphyrin (FTSPP) significantly reduced porphyrin toxicity and inhibit gram-negative bacterial colonization (Sułek et al. 2019). In recent, the development of advanced biocidal multifunctional textiles with a nanocomposite of TiO_2 and Ag nanoparticles is useful against bacterial growth during surgery and accident (Li et al. 2017) to prevent uncontrol bleeding and microbial infections.

CONCLUSION

Here, we have discussed the advances of TiO_2 in environmental remediation as well as in biomedical applications. Despite numerous significant properties, the photocatalytic property of TiO_2 makes it an exceptional nanomaterial among others. Because of photocatalyst, it ideally functions as an antibacterial activity by producing several reactive oxygen species under light irradiation. The photocatalytic activity of TiO_2 is beneficial for environmental remediation purposes in the mitigation of several persistent organic pesticides and antibiotics dissolved in an aquatic environment. Besides this, it is extensively applied in health care devices for the treatment of severe and dreaded diseases like cancer, orthopedic implant, dental, etc. The major question arises of target drug delivery and post-operation infections. Both issues are resolved by the revolutionary applications with biocompatibility, nontoxicity, photocatalytic, stability, high strength, and many more discussed properties of TiO_2.

Despite rigorous work happened on the synthesis of green technology, TiO$_2$, several more qualitative and effective investigations are required to provide direct benefit to the community. Doping of TiO$_2$ with metal, non-metal, nanoparticles and other semiconductors-based photocatalytic processes have been modified and used at a laboratory scale only. However, in the future, the focus should be on the optimization of the system, enhancement in the efficiency of photocatalyst, and emphasize its deployment at pilot scale and benefited to the society. Wastewater treatment is a major task for industries because it consumes a large amount of energy during the process which could be another challenge. There should be the development of a source of light that consumes less energy and can be easily used by the industries to degrade pesticides and other prominent contaminants in water surfaces because UV light has a high cost. With increasing the risk of cancer, the development of new technologies with controllable and minimum side effects causing therapy and diagnosis tools are required. Like TiO$_2$ Nanoparticles, several other nanoparticles with better combinations to overcome the present limitations are urgently required in the future.

REFERENCES

Abdennouri, M., Baâlala, M., Galadi, A., El Makhfouk, M., Bensitel, M., Nohair, K., Sadiq, M., Boussaoud, A. and Barka, N., 2016. Photocatalytic degradation of pesticides by titanium dioxide and titanium pillared purified clays. *Arabian Journal of Chemistry*, 9, pp.S313-S318.

Akakuru, O. U., Iqbal, Z. M. and Wu, A., 2020. TiO2 nanoparticles: properties and applications. TiO2 Nanoparticles: *Applications in Nanobiotechnology and Nanomedicine*, pp.1-66.

Albukhaty, S., Al-Bayati, L., Al-Karagoly, H. and Al-Musawi, S., 2020. Preparation and characterization of titanium dioxide nanoparticles and in vitro investigation of their cytotoxicity and antibacterial activity against *Staphylococcus aureus* and *Escherichia coli*. *Animal Biotechnology*, pp.1-7.

Amalraj, A. and Pius, A., 2015. Photocatalytic degradation of monocrotophos and chlorpyrifos in aqueous solution using TiO_2 under UV radiation. *Journal of Water Process Engineering, 7*, pp.94-101.

Awad, N. K., Edwards, S. L. and Morsi, Y. S., 2017. A review of TiO_2 NTs on Ti metal: Electrochemical synthesis, functionalization and potential use as bone implants. *Materials Science and Engineering: C, 76*, pp.1401-1412.

Behranvand, N., Nasri, F., Emmameh, Z., Khani, P., Hosseini, A., Garssen, J. and Falak, R., 2021. Chemotherapy: a double-edged sword in cancer treatment. *Cancer Immunology, Immunotherapy*, pp.1-20.

Bi, H., Ma, S., Li, Q. and Han, X., 2016. Magnetically triggered drug release from biocompatible microcapsules for potential cancer therapeutics. *Journal of Materials Chemistry B, 4(19)*, pp.3269-3277.

Bi, H., Ma, S., Li, Q. and Han, X., 2016. Magnetically triggered drug release from biocompatible microcapsules for potential cancer therapeutics. *Journal of Materials Chemistry B, 4(19)*, pp.3269-3277.

Blatt, S., Pabst, A. M., Schiegnitz, E., Hosang, M., Ziebart, T., Walter, C., Al-Nawas, B. and Klein, M. O., 2018. Early cell response of osteogenic cells on differently modified implant surfaces: Sequences of cell proliferation, adherence and differentiation. *Journal of Cranio-Maxillofacial Surgery, 46(3)*, pp.453-460.

Çakıroğlu, B. and Özacar, M., 2018. A self-powered photoelectrochemical glucose biosensor based on supercapacitor Co3O4-CNT hybrid on TiO_2. *Biosensors and Bioelectronics, 119*, pp.34-41.

Çeşmeli, S. and Biray Avci, C., 2019. Application of titanium dioxide (TiO2) nanoparticles in cancer therapies. *Journal of drug targeting, 27(7)*, pp.762-766.

Chiang, H. J., Chou, H. H., Ou, K. L., Sugiatno, E., Ruslin, M., Waris, R. A., Huang, C. F., Liu, C. M. and Peng, P. W., 2018. Evaluation of surface characteristics and hemocompatibility on the oxygen plasma-modified biomedical titanium. *Metals, 8(7)*, p.513.

Cloutier, M., Mantovani, D. and Rosei, F., 2015. Antibacterial coatings: challenges, perspectives, and opportunities. *Trends in biotechnology, 33*(11), pp.637-652.

Darvishi, M. H., Nomani, A., Hashemzadeh, H., Amini, M., Shokrgozar, M. A. and Dinarvand, R., 2017. Targeted DNA delivery to cancer cells using a biotinylated chitosan carrier. *Biotechnology and applied biochemistry, 64*(3), pp.423-432.

Das, R., Ambardekar, V. and Bandyopadhyay, P. P., 2021. Titanium Dioxide and Its Applications in Mechanical, Electrical, Optical, and Biomedical Fields. *Titanium Dioxide.*

De Breij, A., Riool, M., Kwakman, P. H. S., De Boer, L., Cordfunke, R.A., Drijfhout, J. W., Cohen, O., Emanuel, N., Zaat, S. A. J., Nibbering, P. H. and Moriarty, T. F., 2016. Prevention of Staphylococcus aureus biomaterial-associated infections using a polymer-lipid coating containing the antimicrobial peptide OP-145. *Journal of Controlled Release, 222*, pp.1-8.

Du, Y., Ren, W., Li, Y., Zhang, Q., Zeng, L., Chi, C., Wu, A. and Tian, J., 2015. The enhanced chemotherapeutic effects of doxorubicin loaded PEG coated TiO_2 nanocarriers in an orthotopic breast tumor bearing mouse model. *Journal of Materials Chemistry B, 3*(8), pp.1518-1528.

Esfandyari, J., Shojaedin-Givi, B., Hashemzadeh, H., Mozafari-Nia, M., Vaezi, Z. and Naderi-Manesh, H., 2020. Capture and detection of rare cancer cells in blood by intrinsic fluorescence of a novel functionalized diatom. *Photodiagnosis and photodynamic therapy, 30*, p.101753.

Giwa, A., Yusuf, A., Balogun, H. A., Sambudi, N. S., Bilad, M. R., Adeyemi, I., Chakraborty, S. and Curcio, S., 2021. Recent advances in advanced oxidation processes for removal of contaminants from water: A comprehensive review. *Process Safety and Environmental Protection, 146*, pp.220-256.

Gong, Z., Hu, Y., Gao, F., Quan, L., Liu, T., Gong, T. and Pan, C., 2019. Effects of diameters and crystals of titanium dioxide nanotube arrays on blood compatibility and endothelial cell behaviors. *Colloids and Surfaces B: Biointerfaces*, *184*, p.110521.

Gulati, K., Johnson, L., Karunagaran, R., Findlay, D. and Losic, D., 2016. In situ transformation of chitosan films into microtubular structures on the surface of nanoengineered titanium implants. *Biomacromolecules*, *17(4)*, pp.1261-1271.

Hashemzadeh, H., Allahverdi, A., Sedghi, M., Vaezi, Z., Tohidi Moghadam, T., Rothbauer, M., Fischer, M. B., Ertl, P. and Naderi-Manesh, H., 2020. PDMS nano-modified scaffolds for improvement of stem cells proliferation and differentiation in microfluidic platform. *Nanomaterials*, *10*(4), p.668.

Hur, W. and Yoon, S. K., 2017. Molecular pathogenesis of radiation-induced cell toxicity in stem cells. *International journal of molecular sciences*, *18(12)*, p.2749.

Jayasree, A., Ivanovski, S. and Gulati, K., 2021. ON or OFF: Triggered therapies from anodized nano-engineered titanium implants. *Journal of Controlled Release*.

Jia, H. and Kerr, L. L., 2015. Kinetics of drug release from drug carrier of polymer/TiO$_2$ nanotubes composite—p H dependent study. *Journal of Applied Polymer Science*, *132*(7).

Jiang, L., Yao, H., Luo, X., Zou, D., Han, C., Tang, C., He, Y., Yang, P., Chen, J., Zhao, A. and Huang, N., 2020. Copper-mediated synergistic catalytic titanium dioxide nanofilm with nitric oxide generation and anti-protein fouling for enhanced hemocompatibility and inflammatory modulation. *Applied Materials Today*, *20*, p.100663.

Junkar, I., Kulkarni, M., Benčina, M., Kovač, J., Mrak-Poljšak, K., Lakota, K., Sodin-Šemrl, S., Mozetic, M. and Iglič, A., 2020. Titanium dioxide nanotube arrays for cardiovascular stent applications. *ACS omega*, *5(13)*, pp.7280-7289.

Kadian, S., Arya, B. D., Kumar, S., Sharma, S. N., Chauhan, R. P., Srivastava, A., Chandra, P. and Singh, S. P., 2018. Synthesis and Application of PHT - TiO$_2$ Nanohybrid for Amperometric Glucose

Detection in Human Saliva Sample. *Electroanalysis, 30(11)*, pp.2793-2802.

Kanan, S., Moyet, M. A., Arthur, R. B. and Patterson, H. H., 2020. Recent advances on TiO_2-based photocatalysts toward the degradation of pesticides and major organic pollutants from water bodies. *Catalysis Reviews,* 62(1), pp.1-65.

Khoshroo, K., Kashi, T. S. J., Moztarzadeh, F., Tahriri, M., Jazayeri, H. E. and Tayebi, L., 2017. Development of 3D PCL microsphere/TiO_2 nanotube composite scaffolds for bone tissue engineering. *Materials Science and Engineering: C, 70,* pp.586-598.

Kim, J. R. and Kan, E., 2016. Heterogeneous photocatalytic degradation of sulfamethoxazole in water using a biochar-supported TiO2 photocatalyst. *Journal of environmental management, 180,* pp.94-101.

Kou, J., Dou, D. and Yang, L., 2017. Porphyrin photosensitizers in photodynamic therapy and its applications. *Oncotarget,* 8(46), p.81591.

Kovalakova, P., Cizmas, L., McDonald, T. J., Marsalek, B., Feng, M. and Sharma, V. K., 2020. Occurrence and toxicity of antibiotics in the aquatic environment: A review. *Chemosphere, 251,* p.126351.

Kumar, A., Khan, M., He, J. and Lo, I. M., 2020. Recent developments and challenges in practical application of visible–light–driven TiO2–based heterojunctions for PPCP degradation: a critical review. *Water research, 170,* p.115356.

Kutuzova, A., Dontsova, T. and Kwapinski, W., 2021. Application of TiO_2-based Photocatalysts to Antibiotics Degradation: Cases of Sulfamethoxazole, Trimethoprim and Ciprofloxacin. *Catalysts, 11*(6), p.728.

Kwatra, B. and Mudgil, M., 2020. Light Assisted Tio2-Based Nanocomposite Systems: A Novel Treatment for Cancer. *Int. J. Med. Biomed. Stud, 4,* pp.28-32.

Lai, M., Jin, Z. and Su, Z., 2017. Surface modification of TiO2 nanotubes with osteogenic growth peptide to enhance osteoblast differentiation. *Materials Science and Engineering: C, 73,* pp.490-497.

Li, S., Zhu, T., Huang, J., Guo, Q., Chen, G. and Lai, Y., 2017. Durable antibacterial and UV-protective Ag/TiO$_2$@ fabrics for sustainable biomedical application. *International journal of nanomedicine, 12*, p.2593.

Li, X., Liu, P., Mao, Y., Xing, M. and Zhang, J., 2015. Preparation of homogeneous nitrogen-doped mesoporous TiO$_2$ spheres with enhanced visible-light photocatalysis. *Applied Catalysis B: Environmental, 164*, pp.352-359.

Liou, T. H., Hung, L. W., Liu, C.L. and Zhang, T. Y., 2018. Direct synthesis of nano titania on highly-ordered mesoporous SBA-15 framework for enhancing adsorption and photocatalytic activity. *Journal of Porous Materials, 25*(5), pp.1337-1347.

Liu, Y., Xie, C., Zhang, F. and Xiao, X., 2019. pH-responsive TiO$_2$ nanotube drug delivery system based on iron coordination. *Journal of Nanomaterials*, 2019.

Luna-Sanguino, G., Ruíz-Delgado, A., Tolosana-Moranchel, A., Pascual, L., Malato, S., Bahamonde, A. and Faraldos, M., 2020. Solar photocatalytic degradation of pesticides over TiO$_2$-rGO nanocomposites at pilot plant scale. *Science of The Total Environment, 737*, p.140286.

Mahboob, S., Nivetha, R., Gopinath, K., Balalakshmi, C., Al-Ghanim, K. A., Al-Misned, F., Ahmed, Z. and Govindarajan, M., 2021. Facile synthesis of gold and platinum doped titanium oxide nanoparticles for antibacterial and photocatalytic activity: A photodynamic approach. *Photodiagnosis and Photodynamic Therapy, 33*, p.102148.

Manivasagam, V. K., Sabino, R. M., Kantam, P. and Popat, K., 2021. Surface Modification Strategies to Improve Titanium Hemocompatibility: A Comprehensive Review. *Materials Advances.*

Mavrič, T., Benčina, M., Imani, R., Junkar, I., Valant, M., Kralj-Iglič, V. and Iglič, A., 2018. Electrochemical biosensor based on TiO$_2$ nanomaterials for cancer diagnostics. In *Advances in Biomembranes and Lipid Self-Assembly* (Vol. 27, pp. 63-105). Academic Press.

Nam, Y., Lim, J. H., Ko, K. C. and Lee, J. Y., 2019. Photocatalytic activity of TiO$_2$ nanoparticles: a theoretical aspect. *Journal of Materials Chemistry A, 7(23)*, pp.13833-13859.

Pan, C., Liu, T., Yang, Y., Liu, T., Gong, Z., Wei, Y., Quan, L., Yang, Z. and Liu, S., 2020. Incorporation of Sr2+ and Ag nanoparticles into TiO$_2$ nanotubes to synergistically enhance osteogenic and antibacterial activities for bone repair. *Materials & Design, 196*, p.109086.

Podporska-Carroll, J., Panaitescu, E., Quilty, B., Wang, L., Menon, L. and Pillai, S. C., 2015. Antimicrobial properties of highly efficient photocatalytic TiO$_2$ nanotubes. *Applied Catalysis B: Environmental, 176*, pp.70-75.

Raines, A. L., Berger, M. B., Schwartz, Z. and Boyan, B. D., 2019. Osteoblasts grown on microroughened titanium surfaces regulate angiogenic growth factor production through specific integrin receptors. *Acta biomaterialia, 97*, pp.578-586.

Razavi, H. and Janfaza, S., 2015. Ethosome: A nanocarrier for transdermal drug delivery. *Archives of Advances in Biosciences, 6(2)*, pp.38-43.

Rehman, F. U., Zhao, C., Jiang, H., Selke, M. and Wang, X., 2016. Photoactivated TiO$_2$ nanowhiskers and tetra sulphonatophenyl porphyrin normoglycemic effect on diabetes mellitus during photodynamic therapy. *Journal of Nanoscience and Nanotechnology, 16(12)*, pp.12691-12694.

Roguska, A., Belcarz, A., Zalewska, J., Hołdyński, M., Andrzejczuk, M., Pisarek, M. and Ginalska, G., 2018. Metal TiO$_2$ nanotube layers for the treatment of dental implant infections. *ACS applied materials & interfaces, 10(20)*, pp.17089-17099.

Safavipour, M., Kharaziha, M., Amjadi, E., Karimzadeh, F. and Allafchian, A., 2020. TiO$_2$ nanotubes/reduced GO nanoparticles for sensitive detection of breast cancer cells and photothermal performance. *Talanta, 208*, p.120369.

Sánchez-López, E., Gomes, D., Esteruelas, G., Bonilla, L., Lopez-Machado, A. L., Galindo, R., Cano, A., Espina, M., Ettcheto, M.,

Camins, A. and Silva, A. M., 2020. Metal-based nanoparticles as antimicrobial agents: an overview. *Nanomaterials, 10*(2), p.292.

Shetti, N. P., Bukkitgar, S. D., Reddy, K. R., Reddy, C. V. and Aminabhavi, T. M., 2019. Nanostructured titanium oxide hybrids-based electrochemical biosensors for healthcare applications. *Colloids and Surfaces B: Biointerfaces, 178*, pp.385-394.

Sułek, A., Pucelik, B., Kuncewicz, J., Dubin, G. and Dąbrowski, J. M., 2019. Sensitization of TiO_2 by halogenated porphyrin derivatives for visible light biomedical and environmental photocatalysis. *Catalysis Today, 335*, pp.538-549.

Tiwari, A., Shukla, A., Tiwari, D. and Lee, S. M., 2018. Nanocomposite thin films Ag0 (NP)/TiO_2 in the efficient removal of micro-pollutants from aqueous solutions: A case study of tetracycline and sulfamethoxazole removal. *Journal of environmental management, 220*, pp.96-108.

Van Viet, P., Phan, B. T., Mott, D., Maenosono, S., Sang, T. T. and Thi, C. M., 2018. Silver nanoparticle loaded TiO_2 nanotubes with high photocatalytic and antibacterial activity synthesized by photoreduction method. *Journal of Photochemistry and Photobiology A: Chemistry, 352*, pp.106-112.

Varma, K. S., Tayade, R. J., Shah, K. J., Joshi, P. A., Shukla, A. D. and Gandhi, V. G., 2020. Photocatalytic degradation of pharmaceutical and pesticide compounds (PPCs) using doped TiO_2 nanomaterials: A review. *Water-Energy Nexus, 3*, pp.46-61.

Vishnu, J., Calin, M., Pilz, S., Gebert, A., Kaczmarek, B., Michalska-Sionkowska, M., Hoffmann, V. and Manivasagam, G., 2020. Superhydrophilic nanostructured surfaces of beta Ti29Nb alloy for cardiovascular stent applications. *Surface and Coatings Technology, 396*, p.125965.

Vu, C. A. and Chen, W. Y., 2019. Field-effect transistor biosensors for biomedical applications: recent advances and future prospects. *Sensors, 19*(19), p.4214.

Wang, J., Xu, G., Zhang, X., Lv, J., Zhang, X., Zheng, Z. and Wu, Y., 2015. Electrochemical performance and biosensor application of TiO_2 nanotube arrays with mesoporous structures constructed by chemical etching. *Dalton Transactions, 44(16)*, pp.7662-7672.

Wang, Q., Huang, J. Y., Li, H. Q., Zhao, A. Z. J., Wang, Y., Zhang, K. Q., Sun, H. T. and Lai, Y. K., 2017. Recent advances on smart TiO_2 nanotube platforms for sustainable drug delivery applications. *International journal of nanomedicine, 12*, p.151.

Wang, T., Liu, X., Zhu, Y., Cui, Z. D., Yang, X. J., Pan, H., Yeung, K. W. K. and Wu, S., 2017. Metal ion coordination polymer-capped pH-triggered drug release system on titania nanotubes for enhancing self-antibacterial capability of Ti implants. *ACS Biomaterials Science & Engineering, 3(5)*, pp.816-825.

Wetchakun, K., Wetchakun, N. and Sakulsermsuk, S., 2019. An overview of solar/visible light-driven heterogeneous photocatalysis for water purification: TiO_2-and ZnO-based photocatalysts used in suspension photoreactors. *Journal of industrial and engineering chemistry, 71*, pp.19-49.

Yang, H., Yu, M., Wang, R., Li, B., Zhao, X., Hao, Y., Guo, Z. and Han, Y., 2020. Hydrothermally grown TiO_2-nanorods on surface mechanical attrition treated Ti: Improved corrosion fatigue and osteogenesis. *Acta Biomaterialia, 116*, pp.400-414.

Yang, T., Qian, S., Qiao, Y. and Liu, X., 2016. Cytocompatibility and antibacterial activity of titania nanotubes incorporated with gold nanoparticles. *Colloids and Surfaces B: Biointerfaces, 145*, pp.597-606.

Yang, W., Ok, Y. S., Dou, X., Zhang, Y., Yang, M., Wei, D. and Xu, P., 2019. Effectively remediating spiramycin from production wastewater through hydrolyzing its functional groups using solid superacid TiO2/SO4. *Environmental research, 175*, pp.393-401.

Yang, Y., Li, X., Qiu, H., Li, P., Qi, P., Maitz, M. F., You, T., Shen, R., Yang, Z., Tian, W. and Huang, N., 2017. Polydopamine modified TiO₂ nanotube arrays for long-term controlled elution of bivalirudin and improved hemocompatibility. ACS applied materials & interfaces, 10(9), pp.7649-7660.

Yasutake, H., Islam, M. S., Rahman, M. A., Yagyu, J., Fukuda, M., Shudo, Y., Kuroiwa, K., Sekine, Y. and Hayami, S., 2021. High proton conductivity from titanium oxide nanosheets and their variation based on crystal phase. *Bulletin of the Chemical Society of Japan.*

Yu, Y., Ran, Q., Shen, X., Zheng, H. and Cai, K., 2020. Enzyme responsive titanium substrates with antibacterial property and osteo/angio-genic differentiation potentials. *Colloids and Surfaces B: Biointerfaces, 185*, p.110592.

Yuan, Z., Huang, S., Lan, S., Xiong, H., Tao, B., Ding, Y., Liu, Y., Liu, P. and Cai, K., 2018. Surface engineering of titanium implants with enzyme-triggered antibacterial properties and enhanced osseointegration in vivo. *Journal of Materials Chemistry B, 6*(48), pp.8090-8104.

Yurt, F., Ocakoglu, K., Ince, M., Colak, S. G., Er, O., Soylu, H. M., Gunduz, C., Biray Avci, C. and Caliskan Kurt, C., 2018. Photodynamic therapy and nuclear imaging activities of zinc phthalocyanine - integrated TiO₂ nanoparticles in breast and cervical tumors. *Chemical biology & drug design, 91(3)*, pp.789-796.

Zahoor, M., Arshad, A., Khan, Y., Iqbal, M., Bajwa, S. Z., Soomro, R. A., Ahmad, I., Butt, F. K., Iqbal, M. Z., Wu, A. and Khan, W. S., 2018. Enhanced photocatalytic performance of CeO 2–TiO 2 nanocomposite for degradation of crystal violet dye and industrial waste effluent. *Applied Nanoscience, 8(5)*, pp.1091-1099.

Zhang, Y., Hu, K., Xing, X., Zhang, J., Zhang, M. R., Ma, X., Shi, R. and Zhang, L., 2021. Smart Titanium Coating Composed of Antibiotic Conjugated Peptides as an Infection - Responsive Antibacterial Agent. *Macromolecular Bioscience, 21*(1), p.2000194.

Zhao, C., Rehman, F. U., Yang, Y., Li, X., Zhang, D., Jiang, H., Selke, M., Wang, X. and Liu, C., 2015. Bio-imaging and photodynamic therapy with tetra sulphonatophenyl porphyrin (TSPP)-TiO 2 nanowhiskers: new approaches in rheumatoid arthritis theranostics. *Scientific reports, 5(1)*, pp.1-11.

Zhao, J., Xu, J., Jian, X., Xu, J., Gao, Z. and Song, Y. Y., 2020. NIR light-driven photocatalysis on amphiphilic TiO_2 nanotubes for controllable drug release. *ACS applied materials & interfaces, 12(20)*, pp.23606-23616.

In: Titanium Dioxide
Editor: Aparna B. Gunjal
ISBN: 978-1-68507-457-9
© 2022 Nova Science Publishers, Inc.

Chapter 3

DOPED TITANIUM DIOXIDE NANOSTRUCTURES FOR VISIBLE LIGHT-DRIVEN PHOTOCATALYSIS

Seema Maheshwari[1], Kuldeep Kaur[1,],*
Ashok Kumar Malik[2] and Simrat Kaur[1]

[1]Department of Chemistry, Mata Gujri College,
Fatehgarh Sahib, Punjab, India
[2]Department of Chemistry, Punjabi University,
Patiala, Punjab, India

ABSTRACT

Titanium dioxide (TiO_2) is a widely used photocatalyst as a result of its stability, non-toxic nature, low cost, and high efficiency. With a bandgap of nearly 3.2 eV, pure TiO_2 is active in the UV region which limits its visible-light photocatalytic applications. Doping with metal or non-metal elements is known to modify the bandgap of TiO_2 to enable it to absorb in the visible region. Dopants play a vital role in enhancing

---

[*] Corresponding Author's E-mail:shergillkk@gmail.com.

the activity and sensitivity of photocatalysts by modification of their optical and electronic properties. Doped TiO_2 shows enhanced activity due to reduced energy bandgap and better separation of electron and hole pairs. The visible light-sensitive titanium dioxide photocatalyst prepared by doping allows the catalyst to utilize solar energy efficiently to enhance its industrial applications. The variation in synthesis conditions and dopant concentration also influence the physicochemical properties of doped titanium dioxide nanostructures. The dopants also play a vital role in modifying the structural and electronic properties to enhance the efficiency of titanium dioxide nanostructures for photocatalytic applications. Visible light active TiO_2 photocatalysts form promising candidates for several applications including degradation of pollutants, carbon dioxide reduction, and water splitting /hydrogen production applications. In this chapter, we explore the synthesis strategies for dopant incorporation and the effect of metal and non-metal dopants on the structural, electronic, and optical properties of titanium dioxide nanostructures for enhanced performance in photocatalytic applications. The challenges and prospects in the development of highly efficient visible light active TiO_2 photocatalysts are also discussed.

Keywords: titanium dioxide, photocatalyst, photocatalytic efficiency, bandgap, visible light catalyst

INTRODUCTION

Developing efficient photocatalysts suited for industrial applications is quite challenging (Zhang et al., 2019). The desirable features of the photocatalysts suited for industrial applications include high photocatalytic efficiency, appropriate bandgap, high visible light harvest ratio, high electron-hole mobility, and recyclability (Li, 2015; Zhang et al., 2019). A wide range of semiconductor nanomaterials including metal oxides (e.g., TiO_2, Fe_2O_3, ZnO, WO_3, and V_2O_5), nitrides (e.g., GaN, Ge_3N_4), and sulphides (e.g., ZnS, CdS) have been investigated for photocatalytic applications (Rani et al., 2018). Although metal nitrides and sulphides have a smaller bandgap than metal oxides which enables them to harness visible light, they suffer from extensive photo corrosion and poor aqueous stability (Rani et al., 2018). Among semiconductor metal oxides, titanium dioxide (TiO_2), also known as titania is a highly promising material for photocatalytic applications. The outstanding properties of titania which include its non-toxic nature, low cost, high thermal and chemical stability, aqueous stability, and resistance to photo corrosion make it an ideal material for several applications including hydrogen production, carbon dioxide reduction, and pollutant degradation (Kanan et al., 2020). TiO_2 is known to exist in three crystal forms- anatase, rutile, and brookite which differ in their photocatalytic activity (Moma & Baloyi, 2018). Among these phases, the anatase and rutile phases are found to be active with the highest photocatalytic activity shown by the anatase phase. Sometimes, a mixture of anatase and rutile phases is also used. The nanostructured TiO_2 has generated much interest for photocatalytic applications as the reduction from bulk to nanosize results in advantages like high surface area and tunable electronic properties (I. Ali et al., 2018). The nanosize affects the bandgap of titania and the charge separation thus affecting its photocatalytic activity. The TiO_2 nanomaterials of varied morphologies like rods, flowers, and tubes can be synthesized using different synthetic approaches which exhibit different photocatalytic properties (Rani et al., 2018). The TiO_2 photocatalytic technology has significantly benefitted from advances in nanotechnology to find solutions for energy and environment-related problems (I. Ali et al., 2018).

The photocatalytic action of TiO_2 arises due to photogenerated electrons and holes. When the light of a particular wavelength (greater than or equal to the bandgap) falls on TiO_2, the energy of photons is used to excite electrons from the valence band (VB) to the conduction band (CB) which creates holes in the valence band. The photogenerated electrons and holes travel to the surface and participate in photocatalysis reactions. The electrons in CB react with oxygen molecules dissolved in water to form superoxide anion radicals ($O_2^{\cdot-}$) and the holes in the VB are captured by -OH and H_2O to form hydroxide radicals $OH^{\cdot}$. The holes, $OH^{\cdot}$, $O_2^{\cdot-}$ and other reactive species thus produced are responsible for the photocatalysis process (Schneider et al., 2014). The commercial application of bare TiO_2 is limited by its low photocatalytic efficiency which arises due to the fast recombination of photogenerated electron-hole pairs (Ansari and Cho, 2016). Also, bare TiO_2 absorbs mainly in the UV region (bandgap of anatase and brookite, ~3.2 eV, rutile, ~3.0 eV,) which constitutes only 5% of the solar spectrum and has very limited absorption in the visible region. For the practical application of TiO_2 photocatalyst, it is important to enhance the visible light-harvesting ability of TiO_2 by modifying its electronic structure to optimize its optical properties (Li, 2015) (Wang et al., 2015). Doping with metal or non-metal ions is considered to be an effective strategy to modify the properties of titania for enhanced performance in photocatalytic applications. Transition metal dopants (e.g., V, Cr, Mn, Fe, Cu, Co, Ni, Mo, Nb, Au, Ru, Ag, and Pt) and nonmetal dopants (e.g., C, N, S, B, P, and I) have been widely studied for enhancing the photocatalytic activity of TiO_2. The incorporation of metal dopant or nonmetal dopant significantly narrows down the bandgap energy and shifts the bandgap edge towards longer wavelengths. The metal dopants show clear influences on the 3d electrons of titanium to elevate or depress the minima of the conduction band, while the non-metal dopants mainly interact with the 2p electrons of oxygen to change the position of the maxima of the valence band (Varma et al., 2020a). The addition of metal dopant can act as an electron trap and is reported to facilitate charge separation thus, reducing its charge carrier recombination rate, and substitution of oxygen by some non-metal dopant creates oxygen defect (Varma et al., 2020).

In this chapter, we shall discuss the preparation strategies for the incorporation of metal and non-metal dopants into the TiO_2 structure reflecting on the effect of the synthesis conditions on doped TiO_2 properties. The effect of dopant on the photocatalytic properties of TiO_2 for improved performance in several recent applications shall also be covered. Finally, the prospects in photocatalysis by doped TiO_2 nanostructures will be highlighted.

SYNTHESIS METHODS

Doped TiO$_2$ nanostructures can typically be prepared by in situ treatment or post-synthesis treatment of TiO$_2$ nanomaterials with the solution of the doping species, thermal treatments to doping species, high-energy ion implantation, or incorporation of active electrolytic species. For these approaches, various synthetic techniques have been applied such as precipitation, hydrothermal/solvothermal, sol-gel, microwave-assisted method, and solid-state reaction method. The synthesis method greatly affects the particle size, surface area, surface properties, and morphologies of synthesized materials which further affect their photocatalytic activities (Binitha et al., 2010). This section covers some of the commonly used synthetic methodologies for the preparation of doped TiO$_2$ nanoparticles and the effect of the synthetic method on the physicochemical properties of doped TiO$_2$.

Sol–Gel Method

The sol-gel method offers an easy, simple, and cost-effective technique to prepare doped TiO$_2$ nanostructures (T. Ali et al., 2018). The sol-gel method is a wet-chemical technique in which nanostructured TiO$_2$ is formed from its precursors e.g., titanium (IV) alkoxide by hydrolysis. The hydrolysis is followed by condensation, drying, and thermal decomposition. The sol-gel method can produce a wide range of novel and functional materials at room temperature with low production costs (Sakka, 2016). The method does not require high temperatures or special equipment as in some of the other methods (Ismael, 2020a). The method allows control over the morphology and size of the particles along with the doping levels through control over the reaction conditions such as pH, temperature, etc. (Shetty et al., 2017). The method produces homogenous high purity particles suited for photocatalytic applications (Valencia et al., 2013). Ali et al., prepared Ag-doped TiO$_2$ from titanium tetraisopropoxide and silver nitrate by sol-gel method at room temperature followed by calcination at 400°C (T. Ali et al., 2018). The XRD and Raman spectra confirmed the incorporation of dopant into the TiO_2 structure. Crystalline particles of anatase phase with uniform morphology were obtained as revealed from SEM and TEM images. The crystallite size was found to decrease with an increase in doping percentage of Ag from 0 to 2.0 mole % as a result of the restrictions on growth imposed by the dopant. The Ag-doped TiO_2 photocatalyst showed enhanced photocatalytic activity. In another work, Ag-doped TiO_2 films were synthesized on silicon substrate by sol-gel process spin coating method followed by calcination at 500°C (Demirci et al., 2016). Anatase phase titania with non-uniform morphology and cracks on the surface was obtained. The Ag doping (0.1 to 0.9%) was found to affect the crystallite size with larger sizes in some cases and smaller ones in some others. The surface defects were also found to be larger in the case of doped titania with the presence of Ag on the surface. The silver doped titania synthesized by the sol-gel spin coating process was found to have higher photocatalytic efficiency (54.13%) than undoped TiO_2 (36.32%) (Demirci et al., 2016). The sol-gel method has proved successful in producing high purity doped TiO_2 nanostructures. However, some limitations such as high cost of precursors, shrinkage of the gel on drying, and residual porosity may restrict its use for the synthesis of doped TiO_2 nanomaterials.

Hydrothermal/Solvothermal Method

In the hydrothermal or solvothermal method, the precursors of TiO_2 dispersed in water or some solvents are heated at an elevated temperature inside an autoclave or pressure vessel. It is one of the versatile methods for the preparation of particles with narrow crystallite size and dispersity (Sirivallop et al., 2020). The morphology of the product can be altered easily to prepare specific shapes and sizes (Dubey and Singh, 2017). The hydrothermal process is often used for the preparation of TiO_2 nanostructures with enhanced catalytic activity. The synthesis of Cr doped TiO_2 was reported by hydrothermal method using ethanol as a solvent (Dubey and Singh, 2017). The synthesis resulted in uniform pure anatase phase nanoparticles with narrow size distribution in the range of 11-13 nm. The Cr^{3+} ion was considered to occupy the main lattice sites of TiO_2 as evident from the absence of impurity peaks in the XRD spectrum. The size of doped particles was found to be smaller than undoped nanoparticles and the energy gap was found to shift to a higher wavelength. Tabasideh et al., synthesized Fe doped TiO_2 via hydrothermal method using pristine TiO_2 and Fe_2O_3 precursors dissolved in HCl (Tabasideh et al., 2017). The product was obtained by heating in an autoclave at 110°C for 12 h and further calcined at 500°C. The absence of impurity peaks in the XRD spectrum confirmed the maintenance of the crystalline anatase phase on doping. The crystallite size of Fe doped TiO_2 (1.5% Fe) was determined to be smaller (29 nm) than undoped TiO_2 (32 nm) and the bandgap shifted to higher wavelengths. However, the particle size was found to be dependent on doping concentration. The doped nanoparticles of pseudospherical morphology were obtained as evident from SEM images. A novel ultrasonic-assisted hydrothermal method has also been used to synthesize Fe-doped TiO_2 nanoparticles using tetrabutyl titanate as Ti precursor and ferric nitrate as Fe precursor in NaOH solution (Sood et al., 2015). The hydrothermal treatment was carried out at 200°C for 12 h after ultrasonication for 1 h and subsequent annealing was carried out at 450°C. The Fe doped TiO_2 exhibited both anatase and rutile phases, however, the anatase phase was dominant. The method resulted in spherical particles with sizes in the range of 10-20 nm. Also, the bandgap was reduced from 3.2 eV in pure titania to 2.9 eV in doped titania and spectral response extended to 426 nm. Jin et al., synthesized N-doped TiO_2 nanoparticles by hydrothermal method using water as a solvent (Jin et al., 2019). The spherical particles with a narrow size range of 12-20 nm and high specific surface area were produced with a reduced bandgap in doped nanoparticles shifting the absorption to the visible region. However, the costly apparatus, inability to observe the reaction progress, and safety are some of the concerns regarding this method (Ismael, 2020b).

Precipitation Method

Chemical coprecipitation is a low-cost and straightforward method for preparing doped TiO_2 materials on large scale. The method produces particles with small size and high homogeneity (Ismael, 2020b). In some cases, the obtained nanopowder is calcinated. Calcination affects the crystal morphology, structure, and optical properties and may lead to an increase in crystal size (Horti et al., 2019). The synthesis of Fe (III) doped TiO_2 with $TiCl_4$ and iron nitrate as the precursors for Ti and Fe, respectively was reported using the precipitation method (Ismael, 2020c). The synthesis resulted in spherical nanoparticles in the anatase phase with a reduced particle size of 12.4 nm (0.1% Fe) as compared to undoped TiO_2 (21 nm). The higher concentration of Fe leads to crystallite deformation and larger particle sizes. The smaller particle size, larger surface area, and reduced charge recombination lead to the higher photocatalytic activity for doped TiO_2 (0.1% Fe). This method has some disadvantages such as severe agglomeration, difficulty in regulating the size, and poor morphology, but is still preferred due to its simplicity and cost-effectiveness.

Microwave-Assisted Method

The synthesis of doped nanoparticles by the microwave-assisted method is gaining popularity due to its simplicity and fast rate of reaction. In this method, the materials are heated by microwave irradiation instead of thermal heating as used in conventional methods. The molecular orientation originated by the interaction of polar molecules with radiations contributes to an improvement in the probability of collision between molecules, resulting in an increased reaction rate. It is a simple and environmentally sustainable process that utilizes energy in a better way relative to traditional forms like the oil or sand bath method. Moreira et al., synthesized TiO_2 nanoparticles doped with Mn through microwave-assisted synthesis method (Moreira et al., 2020). The synthesis involved dispersal of titanium tetraisopropoxide (Ti precursor) in an aqueous isopropanol solution with manganese acetate as the source of Mn followed by irradiation with microwave radiation. The prepared doped nanoparticles were in the anatase phase and exhibited up to 26% enhancement in catalytic activity as compared to pure TiO_2. The reduced particle sizes, larger surface area, and reduction of band gap by 43% obtained for doped TiO_2 were consistent with higher catalytic activity in the visible region (Moreira et al., 2020). Nonetheless, high-cost equipment, unsuitable scale-up production along with the difficulty in monitoring the reaction progress limits the application of this method in doped TiO_2 synthesis.

Solid-State Reaction Method

The solid-state chemical reaction is a promising method for doped TiO_2 production. Compared to other methods such as microwave or hydrothermal, the solid-state reaction method is very cheap and does not need special equipment or solvent to perform. The precursors are mixed by shakers or by ball milling. If the precursors do not react together at room temperature the precursors are calcined at high temperature. This process is used for the large-scale production of doped nanoparticles with high crystallinity. Ghorbanpour et al., recently used a solid-state reaction for the synthesis of Fe doped TiO_2 using TiO_2 and $FeCl_3$ powders at 700°C for 1 h (Ghorbanpour and Feizi, 2019). The method resulted in Fe-doped TiO_2 nanoparticles in the size range of 43.8 to 45.2 nm while undoped TiO_2 particles were 53 nm in diameter which are larger as compared to particles obtained with other methods. The prepared Fe doped nanocatalyst exhibited a smaller bandgap of 2.8 eV as compared to pure TiO_2 (3.1 eV). The catalyst exhibited high visible light activity.

Some limitations of the method include non-uniform particles, long reaction times, and high calcination temperatures (which can result in larger size particles with a low surface area).

PHOTOCATALYTIC PERFORMANCE OF DOPED TiO_2

Titanium dioxide photocatalysts form promising candidates for several applications including degradation of pollutants, carbon dioxide reduction, and water splitting/hydrogen production. Doping is an effective strategy to improve the efficiency of TiO_2 for visible-light photocatalysis for enhanced performance in photocatalytic applications. The photocatalytic activity is also dependent on crystalline structures and surface properties. Doping can modulate the crystal structures and improve photocatalytic properties. In the case of titania, the anatase phase is more photoactive than the rutile phase. The anatase phase exhibits an indirect bandgap which is smaller than the direct bandgap. Therefore, they exhibit a longer lifetime of charge carriers as compared to the rutile phase (Luttrell et al., 2015). The addition of dopants is also known to inhibit the phase transformation from anatase to rutile as the anatase phase is stabilised by the formation of Ti-O-dopant bond thus improving the photocatalytic activity (Mazierski et al., 2018). Doping is also known to retard grain overgrowth, decrease crystallite sizes, and thus increase the surface area. As a result of high surface area, structural defects on the surface are increased and the lattice structure is modified to a small extent. Modified lattices and surface defects help in band gap tuning and thus improve photocatalytic properties.

Metal Dopants

Many transition metal ions (Fe, Ni, Cr, Cu, Mn, Co, and Zn) and lanthanide metal ions (La, Ce, Pr, Nd, Sm, Eu, Dy, and Gd) have been utilized as dopants for improving photocatalytic efficiency of titanium dioxide (Yadav and Jaiswar, 2017). Metal dopants play a vital role in tuning the electronic structure of TiO_2 through dopant-induced states. The addition of dopant may result in the creation of additional donor or acceptor levels above the valence band, below the conduction band, or between the valence and conduction band (Ismael, 2019). The creation of dopant-aligned levels results in the tuning of the bandgap to shift the absorption wavelength to the visible region thus affecting the visible light photocatalytic activity. The dopants may incorporate in interstitial sites or substitute Ti^{4+} ions at lattice positions which depends mainly on the difference in the size of the Ti^{4+} ion and the dopant ion (Ismael, 2019). This substitution results in excess of negative charge that is ionically compensated by the formation of intrinsic defects such as oxygen vacancies, which leads to lowering of absorption energy (Huang et al., 2016). The dopant-induced levels may cause the trapping of electrons, which assists in the separation of charge carriers and reduce the recombination rate by interfacial charge transfer. For example, the doping of Fe^{3+} in TiO_2 resulted in the shift in bandgap energy from 3 eV to 2.1 eV (Moradi et al., 2016a). As a result, the absorption in the visible region increased. The Fe dopant was considered to induce a trapping level between the valence band and conduction band which reduced the recombination rate. The Fe doped titania catalyst showed enhanced photocatalytic activity towards the degradation of dye in the presence of visible light. In Titanium (Ti^{3+}) doped TiO_2, the dopant creates oxygen vacancies and also acts as an electron trapping site shifting the absorption edge to visible region and prolonging the lifetime of electron hole pairs (Wang et al., 2015). The Ti^{3+} doped TiO_2 showed 18.3 times higher activity for photocatalytic degradation of methylene blue as compared to undoped titania. In another work, highly efficient Ru doped TiO_2 visible light active photocatalyst was reported for hydrogen production (Ismael, 2019). The 0.1% doped TiO_2 produced more than twice (3400 μmol/h) hydrogen than undoped TiO_2 (1500 μmol/h). The Ru dopant was considered to improve the visible light photocatalytic activity by introducing donor/acceptor dopant levels and defects in the TiO_2 lattice.

The dopant concentration plays an important role in improving the efficiency of the TiO_2 photocatalyst. The photocatalytic efficiency of titanium dioxide is enhanced on an increase in dopant concentration up to a certain level beyond which the photocatalytic efficiency is found to decrease with an increase in dopant concentration. The increase in photocatalytic activity is concomitant with a decrease in the particle size and an increase in the surface area (Murashkina et al., 2015). In the case of Ru doped TiO_2, the particle size was found to decrease with an increase in the concentration of dopant from 0.05 mol% Ru (18.7 nm) to 0.1 mol% Ru (17.5 nm) while it was found to increase on further increase in dopant concentration from 0.15 mol% Ru (20.4 nm) to 0.2 mol% (19.3 nm) as compared to undoped TiO_2 (21 nm) (Ismael, 2019). The BET surface areas were also found to increase in a similar manner reaching the optimum value at 0.1 mol% Ru. The 0.1 mol% Ru doped TiO_2 was found to possess the highest photocatalytic activity towards hydrogen production. The introduction of Ru^{4+} which has a larger ionic radius (62 pm) than that of Ti^{4+} (60.5 pm) in the crystal structure induced lattice strain and structural distortions. The smaller size and higher surface area in the case of 0.1 mol% Ru doped TiO_2 resulted in higher photocatalytic activity as the electron-hole pairs reached the surface in smaller time which reduced the recombination rate. With an increase in the concentration of dopant, larger lattice distortion is produced which causes an increase in the size of the particles. Similarly, in the case of Cu-doped TiO_2 photocatalyst prepared by sol-gel method, the dopant concentration was found to affect the particle size and surface area (Bhattacharyya et al., 2021). The 1% Cu doped TiO_2 with the smallest particle size and high surface area exhibited the maximum photocatalytic activity for reduction of CO_2 to CH_4 (1081 µL/h/g). The different Cu species (Cu^0, Cu^{1+}, and Cu^{2+}) and oxygen vacancies created as a result of doping were found to play an important role in enhancing photocatalytic activity. In work carried out by Moradi et al. on Fe doped TiO_2 nanoparticles, the particle size was found to decrease with an increase in doping concentration from 0 (23 nm) to 1 wt % (15 nm) with a corresponding increase in surface area (Moradi et al., 2016b). The 1 wt% Fe doped TiO_2 nano-particles

(1wt%) exhibited high photocatalytic activity towards degradation of reactive red 198 (RR 198) under UV and visible light irradiations. The absorption edge of Fe doped TiO_2 shifted to longer wavelengths due to the introduction of new energy levels by the dopant into the bandgap of TiO_2 (Moradi et al., 2016b). In a work involving the synthesis of W^{6+} doped TiO_2 and Mo^{6+} doped TiO_2, the addition of dopant (W or Mo) was associated with smaller crystallite size and higher surface area. The specific surface area of TiO_2 was increased from 144 m^2g^{-1} (pure TiO_2) to 163 m^2g^{-1} and 170 m^2g^{-1} on Mo and W doping, respectively. The doped TiO_2 exhibited 3 times higher catalytic activity than undoped TiO_2 for degradation of 4-chlorophenol.

Lanthanide metals mostly in trivalent oxidation state show great potential as TiO_2 dopants. A wide bandgap TiO_2 with lanthanide dopant having ladder-like energy levels embedded in an appropriate TiO_2 lattice can result in the absorption of light from the near-infrared and ultraviolet wavelengths to the visible spectral range (Mazierski 2018). Additionally, lanthanide-doped TiO_2 usually present luminescent behavior. This luminescence arises due to f–f electronic transitions within their partially filled 4f orbitals (Avram et al., 2021). A study reported highly efficient Er-doped TiO_2 for the degradation of volatile organic compounds including acetaldehyde, o-xylene, and ethylene (Rao et al., 2019). The 1.5% doped sample was found to show maximum degradation efficiency for acetaldehyde. The reasons for increased efficiency were enhanced adsorption capacity (3.3 times that of undoped TiO_2), decreased particle size (from 24.4 nm for undoped to 12.5 nm for doped particles), an increase in specific surface area from 59.28 to 110.34 m^2/g. The Er dopant introduced oxygen vacancies and Ti^{3+} ions into the crystal lattice successfully enhancing its visible light activity and photocatalytic performance. A summary of the visible light photocatalysts based on metal-doped TiO_2 is given in Table 1.

Table 1. Visible light photocatalysts based on metal-doped TiO$_2$

S. No.	Dopant	Synthesis method	Application	% Degradation	Ref. No.
1.	Fe-doped	Sol-gel	Methyl orange	98% in 60 mins	Moradi et al., 2018
2.	Fe-doped	Precipitation method	Methyl orange and 4-CP	95% 65%	Ismael, 2020a
3.	Fe-doped	Hydrothermal method	Diazinon	85%	Tabasideh et al., 2017
4.	Fe-doped	Ultrasonic assisted hydrothermal method	p-Nitrophenol	92% in 5h	Sood et al., 2015
5.	Cu-N co-doped	Sol-gel	Methylene Blue p-Nitro phenol	100% 45% after 2 hr	Jaiswal et al., 2015
6.	Ag-doped	Sol-gel	Methylene Blue	96%	T. Ali et al., 2018
7.	Cu-doped	Sol gel	Bacterial reduction	99.99% in 30 min	Mathew et al., 2018
8.	Ag-doped	Sol gel spin coating	Methylene Blue	54.13%	Demirci et al., 2016
9.	Ru doped	Precipitation Method	Hydrogen production	3400 µmol/h	Ismael, 2019
10.	W or Mo doped TiO$_2$	Evaporation induced self-assembly	4-Chlorophenol	95%	Avilés-García et al., 2017
11.	Er	Sol-gel method	Acetaldehyde o-Xylene	99.2 & 84.6% in 100 min	Rao et al., 2019
12.	Nd	Sol-gel method	Methylene Blue Congo red	92% in 45 mins	Nithya et al., 2018
13.	La-N codoped	Microwave assisted Sol-gel	Naphthalene	93.5% in 120 mins	Liu et al., 2016
14.	Ag-Ce co doped	Microwave assisted chemical reduction	Hydrogen production	1.47 µmol/cm^2/h in 6h	Fan et al., 2015

Table 2. Visible light photocatalysts based on nonmetal doped-TiO$_2$

S. No.	Dopant	Synthesis method	Application	% Degradation	Ref.
1.	N-doped	Sol-gel	Methyl Orange	90% in 200 min	Bergamonti et al., 2017
2.	N-doped	Sol-gel	Furfural	78% in 75 mins	Veisi et al., 2016
3.	N-doped	Sol-gel	RhB	90% in 40 mins	Huang et al., 2017
4.	N-Cu co-doped TiO$_2$	Sol-gel/hydrothermal	COD & TOC	93% & 89% within 180 mins	Isari et al., 2020
5.	N-doped	Sol-gel	Ciprofloxacin	95% in 30 mins	Shetty et al., 2017
6.	N-Ag co-doped	Solvothermal	Methylene Blue NH$_3$ removal	98.82% 37.5%	Sirivallop et al., 2020
7.	S-doped	Hydrothermal method	Dimethyl Sulphide	98% in 120 mins	Lin et al., 2016
8.	S-doped	Hydrothermal method	RhB	80% in 60 mins	Wang et al., 2017
9.	Hydrogenated F-doped	Precipitation	H$_2$ production & MB degradation	100%	Gao et al., 2019
10.	B and Y co-doped TiO$_2$	Sol-gel	Methylene Blue	55%	Wu et al., 2017

Non-Metal Dopants

Non-metals dopants have also been investigated for developing visible-light active TiO_2 photocatalysts (Varma et al., 2020b; Basavarajappa et al., 2020). Some non-metal dopants employed for TiO_2 doping include C, N, I, B, and S. The non-metal dopants substitute oxygen on the surface of TiO_2 thus creating oxygen vacancies. Non-metal dopants alter the bandgap by creating a new hybrid orbital (hybridised by 2p state of N/C/B/S and O 2p states) over the VB of TiO_2 (Gao et al., 2019). The most commonly used non-metal dopant for TiO_2 doping is nitrogen. Various nitrogen precursors such as urea, HNO_3, ammonium hydroxide, and ammonium nitrate have been employed for doping. The N-doped TiO_2 synthesized by sol-gel method with ammonium nitrate as N precursor produced mesoporous TiO_2 with a small average pore size (3.794 nm) and high specific surface area (182.396 m^2/g) (Zhao et al., 2015). The particle sizes of N doped TiO_2 were found to be smaller than undoped TiO_2. The nitrogen doping facilitated the transformation of amorphous TiO_2 to the anatase phase and the absorption edge was shifted to the visible region. The photocatalytic activity towards degradation of rhodamine B was enhanced about 3 times as compared to pure TiO_2. Jin et al. synthesized N-TiO_2 nanoparticles by a hydrothermal method (Jin et al., 2019). The spherical particles with sizes in the range of 12 nm to 20 nm and a specific surface area of 148.52 m^2/g were formed. The reduced bandgap (2.92 eV) shifted the absorption to the visible region. The photocatalyst was able to achieve degradation of norfloxacin up to 99.53% within 30 min. Incorporation of S can substitute both anion and cation in a TiO_2 lattice thus affecting the crystalline and electronic properties of TiO_2. Ramacharyulu et al., reported substitution by sulphide ions in the lattice position of oxygen which lead to a reduction in the bandgap energy of TiO_2 (Ramacharyulu et al., 2015). A summary of some visible light photocatalysts based on non-metal doped TiO_2 is provided in Table 2.

Metal and Non-Metal Co-Doping

Co-doping involving metal and non-metal species has also been proposed as an effective strategy to enhance photocatalytic efficiency through synergistic effects of both the dopants. Cu and N co-doped photocatalyst which possessed photocatalytic activity higher than the mono doped (Cu or N) TiO_2 was obtained (Jaiswal et al., 2015). In a co-doped photocatalyst both the dopant ions play an important role in enhancing efficiency. For example, in N and Ag codoped TiO_2, N-doping enhanced homogenous morphology and specific surface area, while Ag doping reduced the bandgap. The photocatalytic efficiency of N and Ag co-doped TiO_2 was found to be 2.7 and 4.3 times higher for degradation of methylene blue and removal of ammonia respectively as compared to undoped TiO_2 (Sirivallop et al., 2020). La-N co-doped TiO_2 was synthesized by the microwave-assisted sol-gel method. The incorporation of La and N in the TiO_2 framework reduced the bandgap from 2.82 eV to 2.2.0 eV shifting it to the visible region. The N and Ag co-doped catalyst exhibited a high degradation efficiency of 93.5% for degradation of naphthalene which arose as a result of the synergistic effect of both N and Ag which lead to reduced recombination rate, increase in surface area, and enhanced visible-light harvest (Liu et al., 2016). Y. Wu et al. synthesized B and Y co-doped TiO_2 by sol-gel method followed by calcination (Wu et al., 2017). The prepared co-doped photocatalyst exhibited twice the photocatalytic efficiency of undoped TiO_2 towards methylene blue dye. The enhancement was attributed to the Y^{3+} and B^{3+} ions which were incorporated in the lattice of TiO_2 through substitution of lattice positions and occupancy of interstitial sites respectively. The B and Y dopants created structural defects and trap sites for electrons which prolonged carrier lifetimes. The smaller crystallite sizes and enhanced surface-to-volume ratio contributed towards improving the catalytic activity.

CONCLUSION

Doping has proved to be an effective approach to synthesize visible-light-driven photocatalysts suited for several applications including hydrogen production, pollutant degradation, and wastewater purification. The dopant incorporation in the crystal lattice of titanium dioxide results in altered band gaps and properties of charge carriers to affect the photocatalytic activity. The synthesis methods and doping moieties affect the physicochemical characteristics of the synthesized products. The doped TiO_2 nanostructures have been exploited for energy and environment-related problems, however, the large-scale industrial application is still hindered by low photoactivity and low spectral sensitivity to visible and NIR regions. Thus there is a need to modify synthetic routes and explore doping entities to produce TiO_2 nanostructures with still higher photocatalytic activity under visible light irradiation. Also, a thorough understanding of the photocatalysis pathways and mechanisms may help in overcoming the challenges for developing high-efficiency doped TiO_2 photocatalysts.

REFERENCES

Ali, I., Suhail, M., Alothman, Z. A., Alwarthan, A., 2018. Recent advances in syntheses, properties and applications of TiO2 nanostructures. *RSC Adv.* 8, 30125–30147.

Ali, T., Ahmed, A., Alam, U., Uddin, I., Tripathi, P., Muneer, M., 2018. Enhanced photocatalytic and antibacterial activities of Ag-doped TiO_2 nanoparticles under visible light. *Materials Chemistry and Physics* 212, 325–335.

Ansari, S. A., Cho, M. H., 2016. Highly Visible Light Responsive, Narrow Band gap TiO_2 Nanoparticles Modified by Elemental Red Phosphorus for Photocatalysis and Photoelectrochemical Applications. *Sci Rep* 6, 25405.

Avilés-García, O., Espino-Valencia, J., Romero, R., Rico-Cerda, J. L., Arroyo-Albiter, M., & Natividad, R. (2017). W and Mo doped TiO_2: Synthesis, characterization and photocatalytic activity. *Fuel, 198*, 31–41.

Avram, D., Cojocaru, B., & Tiseanu, C. (2021). First evidence from luminescence of lanthanide substitution in rutile TiO_2. *Materials Research Bulletin, 134*, 111091.

Basavarajappa, P. S., Patil, S. B., Ganganagappa, N., Reddy, K. R., Raghu, A. V., & Reddy, Ch. V. (2020). Recent progress in metal-doped TiO_2, non-metal doped/codoped TiO2 and TiO_2 nanostructured hybrids for enhanced photocatalysis. *International Journal of Hydrogen Energy, 45*(13), 7764–7778.

Bergamonti, L., Predieri, G., Paz, Y., Fornasini, L., Lottici, P. P., & Bondioli, F. (2017). Enhanced self-cleaning properties of N-doped TiO_2 coating for Cultural Heritage. *Microchemical Journal, 133*.

Bhattacharyya, K., Mane, G. P., Rane, V., Tripathi, A. K., Tyagi, A. K., 2021. Selective CO_2 Photoreduction with Cu-Doped TiO_2 Photocatalyst: Delineating the Crucial Role of Cu-Oxidation State and Oxygen Vacancies. *J. Phys. Chem. C* 125, 1793–1810.

Binitha, N. N., Yaakob, Z., Resmi, R., 2010. Influence of synthesis methods on zirconium doped titania photocatalysts. *Cent. Eur. J. Chem.* 8, 182–187.

Demirci, S., Dikici, T., Yurddaskal, M., Gultekin, S., Toparli, M., Celik, E., 2016. Synthesis and characterization of Ag doped TiO_2 heterojunction films and their photocatalytic performances. *Applied Surface Science* 390, 591–601.

Dubey, R. S., Singh, S., 2017. Investigation of structural and optical properties of pure and chromium doped TiO_2 nanoparticles prepared by solvothermal method. *Results in Physics* 7, 1283–1288.

Fan, X., Wan, J., Liu, E., Sun, L., Hu, Y., Li, H., Hu, X., Fan, J., 2015. High-efficiency photoelectrocatalytic hydrogen generation enabled by Ag deposited and Ce doped TiO_2 nanotube arrays. *Ceramics International 3 Part B*, 5107–5116.

Gao, Q., Si, F., Zhang, S., Fang, Y., Chen, X., Yang, S., 2019. Hydrogenated F-doped TiO_2 for photocatalytic hydrogen evolution and pollutant degradation. *International Journal of Hydrogen Energy* 44, 8011–8019.

Ghorbanpour, M., Feizi, A., 2019. Iron-doped TiO_2 Catalysts with Photocatalytic Activity. *Journal of Water and Environmental Nanotechnology* 4, 60–66.

Horti, N. C., Kamatagi, M., Patil, N., Sanna Kotrappanavar, N., Sannaikar, M. S., & Inamdar, S. (2019). Synthesis and photoluminescence properties of titanium oxide (TiO_2) nanoparticles: Effect of calcination temperature. *Optik, 194*, 163070.

Huang, F., Yan, A., Zhao, H., 2016. *Influences of Doping on Photocatalytic Properties of TiO_2 Photocatalyst, Semiconductor Photocatalysis-Materials, Mechanisms and Applications.* IntechOpen.

Isari, A. A., Hayati, F., Kakavandi, B., Rostami, M., Motevassel, M., & Dehghanifard, E. (2020). N, Cu co-doped TiO_2@functionalized SWCNT photocatalyst coupled with ultrasound and visible-light: An effective sono-photocatalysis process for pharmaceutical wastewaters treatment. *Chemical Engineering Journal, 392*, 123685.

Ismael, M., 2020a. Enhanced photocatalytic hydrogen production and degradation of organic pollutants from Fe (III) doped TiO_2 nanoparticles. *Journal of Environmental Chemical Engineering* 8, 103676.

Ismael, M., 2020b. A review and recent advances in solar-to-hydrogen energy conversion based on photocatalytic water splitting over doped-TiO_2 nanoparticles. *Solar Energy* 211, 522–546.

Ismael, M., 2019. Highly effective ruthenium-doped TiO_2 nanoparticles photocatalyst for visible-light-driven photocatalytic hydrogen production. *New J. Chem.* 43, 9596–9605.

Jaiswal, R., Bharambe, J., Patel, N., Dashora, A., Kothari, D. C., & Miotello, A. (2015). Copper and Nitrogen co-doped TiO_2 photocatalyst with enhanced optical absorption and catalytic activity. *Applied Catalysis B: Environmental, 168–169*, 333–341.

Jin, X., Zhou, X., Sun, P., Lin, S., Cao, W., Li, Z., Liu, W., 2019. Photocatalytic degradation of norfloxacin using N-doped TiO_2: Optimization, mechanism, identification of intermediates and toxicity evaluation. *Chemosphere* 237, 124433.

Kanan, S., Moyet, M. A., Arthur, R. B., & Patterson, H. H. (2020). Recent advances on TiO_2 -based photocatalysts toward the degradation of pesticides and major organic pollutants from water bodies. *Catalysis Reviews*, *62*(1), 1–65.

Li, W., 2015. Influence of electronic structures of doped TiO_2 on their photocatalysis: Influence of electronic structures of doped TiO_2 on their photocatalysis. *Phys. Status Solidi RRL* 9, 10–27.

Lin, Y. H., Hsueh, H. T., Chang, C. W., & Chu, H. (2016). The visible light-driven photodegradation of dimethyl sulfide on S-doped TiO_2: Characterization, kinetics, and reaction pathways. *Applied Catalysis B: Environmental*, *199*, 1–10.

Liu, D., Wu, Z., Tian, F., Ye, B. C., & Tong, Y. (2016). Synthesis of N and La co-doped TiO_2/AC photocatalyst by microwave irradiation for the photocatalytic degradation of naphthalene. *Journal of Alloys and Compounds*, *676*, 489–498.

Luttrell, T., Halpegamage, S., Tao, J., Kramer, A., Sutter, E., Batzill, M., 2015. Why is anatase a better photocatalyst than rutile? - Model studies on epitaxial TiO_2 films. *Sci Rep* 4, 4043.

Mathew, S., Ganguly, P., Rhatigan, S., Kumaravel, V., Byrne, C., Hinder, S. J., Bartlett, J., Nolan, M., & Pillai, S. C. (2018). Cu-Doped TiO_2: Visible Light Assisted Photocatalytic Antimicrobial Activity. *Applied Sciences*, *8*(11), 2067.

Mazierski, P., Mikolajczyk, A., Bajorowicz, B., Malankowska, A., Zaleska-Medynska, A., Nadolna, J., 2018. The role of lanthanides in TiO_2-based photocatalysis: A review. *Applied Catalysis B: Environmental* 233, 301–317.

Moma, J., Baloyi, J., 2018. *Modified Titanium Dioxide for Photocatalytic Applications, Photocatalysts-Applications and Attributes.* IntechOpen.

Moradi, H., Eshaghi, A., Hosseini, S. R., Ghani, K., 2016. Fabrication of Fe-doped TiO_2 nanoparticles and investigation of photocatalytic decolorization of reactive red 198 under visible light irradiation. *Ultrasonics Sonochemistry* 32, 314–319.

Moreira, A. J., Malafatti, J. O. D., Giraldi, T. R., Paris, E. C., Pereira, E. C., de Mendonça, V. R., Mastelaro, V. R., Freschi, G. P. G., 2020. Prozac® photodegradation mediated by Mn-doped TiO_2 nanoparticles: Evaluation of by-products and mechanisms proposal. *Journal of Environmental Chemical Engineering* 8, 104543.

Murashkina, A. A., Murzin, P. D., Rudakova, A. V., Ryabchuk, V. K., Emeline, A. V., Bahnemann, D. W., 2015. Influence of the Dopant Concentration on the Photocatalytic Activity: Al-Doped TiO_2. *J. Phys. Chem. C* 119, 24695–24703.

Nithya, N., Bhoopathi, G., Magesh, G., & Kumar, C. D. N. (2018). Neodymium doped TiO_2 nanoparticles by sol-gel method for antibacterial and photocatalytic activity. *Materials Science in Semiconductor Processing*, *83*, 70–82.

Ramacharyulu, P. V. R. K., Nimbalkar, D. B., Kumar, J. P., Prasad, G. K., & Ke, S. C. (2015). N-doped, S-doped TiO_2 nanocatalysts: Synthesis, characterization and photocatalytic activity in the presence of sunlight. *RSC Advances*, *5*(47), 37096–37101.

Rani, A., Reddy, R., Sharma, U., Mukherjee, P., Mishra, P., Kuila, A., Sim, L. C., Saravanan, P., 2018. A review on the progress of nanostructure materials for energy harnessing and environmental remediation. *J Nanostruct Chem* 8, 255–291.

Rao, Z., Xie, X., Wang, X., Mahmood, A., Tong, S., Ge, M., Sun, J., 2019. Defect Chemistry of Er^{3+}-Doped TiO_2 and Its Photocatalytic Activity for the Degradation of Flowing Gas-Phase VOCs. *J. Phys. Chem. C* 123, 12321–12334.

Sakka, S., 2016. The Outline of Applications of the Sol–Gel Method, in: Klein, L., Aparicio, M., Jitianu, A. (Eds.), *Handbook of Sol-Gel Science and Technology*. Springer International Publishing, Cham, pp. 1–33.

Schneider, J., Matsuoka, M., Takeuchi, M., Zhang, J., Horiuchi, Y., Anpo, M., & Bahnemann, D. W. (2014). Understanding TiO_2 Photocatalysis: Mechanisms and Materials. *Chemical Reviews*, *114*(19), 9919–9986.

Shetty, R., Chavan, V. B., Kulkarni, P. S., Kulkarni, B. D., Kamble, S. P., 2017. Photocatalytic Degradation of Pharmaceuticals Pollutants Using N-Doped TiO_2 Photocatalyst: Identification of CFX Degradation Intermediates. *Indian Chemical Engineer* 59, 177–199.

Sirivallop, A., Areerob, T., Chiarakorn, S., 2020. Enhanced Visible Light Photocatalytic Activity of N and Ag Doped and Co-Doped TiO_2 Synthesized by Using an In-Situ Solvothermal Method for Gas Phase Ammonia Removal. *Catalysts* 10, 251.

Sood, S., Umar, A., Mehta, S. K., Kansal, S. K., 2015. Highly effective Fe-doped TiO_2 nanoparticles photocatalysts for visible-light driven photocatalytic degradation of toxic organic compounds. *Journal of Colloid and Interface Science* 450, 213–223.

Tabasideh, S., Maleki, A., Shahmoradi, B., Ghahremani, E., & Mckay, G. (2017). Sonophotocatalytic degradation of diazinon in aqueous solution using iron-doped TiO_2 nanoparticles. *Separation and Purification Technology*, *189*.

Valencia, S., Vargas, X., Rios, L., Restrepo, G., Marín, J. M., 2013. Sol–gel and low-temperature solvothermal synthesis of photoactive nano-titanium dioxide. *Journal of Photochemistry and Photobiology A: Chemistry* 251, 175–181.

Varma, K. S., Tayade, R. J., Shah, K. J., Joshi, P. A., Shukla, A. D., Gandhi, V. G., 2020. Photocatalytic degradation of pharmaceutical and pesticide compounds (PPCs) using doped TiO_2 nanomaterials: A review. *Water-Energy Nexus* 3, 46–61.

Veisi, F., Zazouli, M., Ebrahimzadeh, M., Charati, J., & Shiralizadeh Dezfuli, A. (2016). Photocatalytic degradation of furfural in aqueous solution by N-doped titanium dioxide nanoparticles. *Environmental Science and Pollution Research*, 23.

Wang, F., Li, F., Zhang, L., Zeng, H., Sun, Y., Zhang, S., & Xu, X. (2017). S-TiO2 with enhanced visible-light photocatalytic activity derived from TiS$_2$ in deionized water. *Materials Research Bulletin, 87*, 20–26.

Wang, X., Li, Y., Liu, X., Gao, S., Huang, B., Dai, Y., 2015. Preparation of Ti^{3+} self-doped TiO$_2$ nanoparticles and their visible light photocatalytic activity. *Chinese Journal of Catalysis* 36, 389–399.

Wu, Y., Gong, Y., Liu, J., Zhang, Z., Xu, Y., Ren, H., Li, C., & Niu, L. (2017). B and Y co-doped TiO$_2$ photocatalyst with enhanced photodegradation efficiency. *Journal of Alloys and Compounds, 695*, 1462–1469.

Yadav, S., & Jaiswar, G. (2017). Review on Undoped/Doped TiO$_2$ Nanomaterial; Synthesis and Photocatalytic and Antimicrobial Activity. *Journal of the Chinese Chemical Society, 64*(1), 103–116.

Zhang, F., Wang, X., Liu, H., Liu, C., Wan, Y., Long, Y., & Cai, Z. (2019). Recent Advances and Applications of Semiconductor Photocatalytic Technology. *Applied Sciences, 9*(12), 2489.

Zhao, Y., Li, H., Bala, H., Chen, J., Zhang, B., Fu, X., & Fu, W. (2015). Synthesis of visible Light Responsive N Doped TiO$_2$ Photocatalyst and its Enhanced Photocatalytic activity. *Current Nanoscience, 11*, 1–1.

In: Titanium Dioxide
Editor: Aparna B. Gunjal
ISBN: 978-1-68507-457-9
© 2022 Nova Science Publishers, Inc.

Chapter 4

BIOSYNTHESIZED TITANIUM DIOXIDE NANOPARTICLES AND THEIR ANTIBACTERIAL EFFICACY

Immanuel J. Suresh[1], Iswareya V. Lakshimi[1] and Shanthipriya Ajmera[2],*

[1]Department of Microbiology, The American College, Madurai, Tamil Nadu, India
[2]Department of Microbiology, Palamuru University, Mahabubnagar, Telangana, India

ABSTRACT

The rising incidence of bacterial diseases and bacterial attacks in several fields, such as food, agriculture, water treatment, and food packaging, has resulted in a continuing hunt for new antibacterial agents. Some bacterial pathogens have evolved resistance mechanisms to existing antimicrobials. To combat bacterial pathogens, there is a need for specific and alternative antibacterial agents. In comparison to

---

* Corresponding Author's E-mail: spmbpu2017@gmail.com.

their bulk materials, nanoparticles have a higher surface area to volume ratio and have improved antibacterial activity. Because of their broad-spectrum antibacterial properties, metal oxide nanoparticles, particularly titanium dioxide (TiO_2) nanoparticles, have attracted a huge attention. The use of TiO_2 in human food, pharmaceuticals, and cosmetics has been approved by the FDA. TiO_2 nanoparticles are ideal for antibacterial applications such as air purification, water purification, and antimicrobial coatings on biomedical devices due to their unique photocatalytic activity and quantum size effects. The antibacterial activity of TiO_2 nanoparticles against both gram-positive and gramme negative bacteria is has been studied. TiO_2 nanotubes of size 20nm showed 97.53% antibacterial activity against *Escherichia coli*. Similarly, TiO_2 nanoparticles synthesized using *Aspergillus niger* were able to reduce the growth of *Pseudomonas aeruginosa, E. coli* and *Klebsiella pneumoniae*. TiO_2 nanoparticles synthesized using *Azadirachta indica* leaves extract as a reducing agent showed antibacterial activity against *Salmonella typhi, E. coli* and *K. pneumoniae*. TiO_2 nanoparticles modified using plant extracts such as *Garcinia zeylanica* enhanced the antibacterial activity. This review provides a brief knowledge on the antibacterial activity of the TiO_2 nanoparticles and evidences that TiO_2 nanoparticles can be developed as a promising eco-friendly antibacterial agents.

Keywords: TiO_2 nanoparticles, plant extract, microbial source, antibacterial activity

INTRODUCTION

Nanoparticles are substances that have at least one of their dimensions in the range of 1-100 nm and connect the bulk material and atomic structure. It is of huge importance due to their properties like small size, large surface and with free bonds, higher reactivity and expression of various beneficial properties in comparison to their bulk materials. Nanoparticles are used for catalysis and also in all fields like medical, packaging, cosmetics, agriculture, photonics, and many more fields. Metal oxide nanoparticles including Titanium dioxide (TiO_2), copper oxide, Zinc oxide, lead oxide and Fe_3O_4 are used in various biological applications (Varghese et al., 2020). Overuse of antibiotics has led to antimicrobial resistance especially among the bacteria. So there is a huge demand for innovative antibacterial compounds that can be used for combating various gram-positive and gram-negative bacterial pathogens. In this regard, metal oxide nanoparticles especially Titanium dioxide nanoparticles (TiO$_2$ NPs) play a major role and is of major consideration as they are less toxic, in expensive and Generally Recognized as Safe (GRAS) substance. It has various special features like easy control and exhibits good resistance to chemical erosion (Hussain et al., 2016).

Titanium dioxide (TiO$_2$) is used to produce important semiconductor nanoparticles which have a variety of applications like biosensing, computer transistors, electrometers, catalysts to optics, chemical sensors and memory schemes. They also play a major role in health-related applications like antimicrobial activity, antioxidant activity, anticancer activity, medical imaging, filters, nanocomposites, hyperthermia of tumours and delivery of drugs (Kalyanasundaram & Prakash, 2015). TiO$_2$ is a thermally stable and biocompatible chemical compound with high photocatalytic activity and has presented good results against bacterial contamination. TiO$_2$ is also known for its long-term stability with high oxidising power and is also a strong photocatalytic material (de Dicastillo et al., 2020).

TiO$_2$ NPs are of major interest due to their variety of applications in cosmetics, pharmaceuticals, paper inks, skin protection from UV and food colorants. They are preferred because of their variety of properties like photocatalyst, non-toxicity and high chemical stability (Ajmal et al., 2019). Titanium dioxide nanoparticles (TiO$_2$ NPs) are one of the major metal oxide nanoparticles studied in the area of antimicrobial applications due to their properties such as bactericidal photocatalytic activity, safety, and self-cleaning properties. It has a wide range of target pathogens including gram-positive bacteria, gram-negative bacteria and some drug-resistant bacteria (de Dicastillo et al., 2020). The antimicrobial properties of TiO$_2$ NPs depend on various factors like shape, size, morphology, source of synthesis and its photocatalytic effect (Hussain et al., 2016).

BIOSYNTHESIS OF TiO_2 NANOPARTICLES

There are various of TiO_2 NPs synthesis with major methods of chemical and biosynthesis method. Since toxicity of nanoparticles is a major concern in its application, scientists are seeking a more greener and eco-friendly way to develop nanoparticles with less toxicity to host (Varghese et al., 2020). The biosynthetic method of nanoparticle also known as the green synthesis method involves the use of plant extracts, microbes including bacteria, fungi and their metabolites and other enzymes for the production of nanoparticles from their bulk material (de Dicastillo et al., 2020).

Chemical method of TiO_2 NPs synthesis may involve the presence of toxic components which is a major drawback in the chemical synthesis method. Whereas in green synthesis method, especially based on plant extracts, as phytochemicals are involved in the production of nanoparticles, there is less toxicity in comparison with chemical nanoparticles (Rosi & Kalyanasundaram, 2018). The chemical method of nanoparticle synthesis may require toxic and costly reagents which may be not safe for the environment and can increase the reactivity and toxicity of nanoparticles (Hussain et al., 2016). Whereas biosynthesis of nanoparticles will be eco-friendly, biodegradable, sustainable, free of contamination and less expensive (Figure 1).

Production of NPs using biological methods from natural substances is emerging as an important area in nanotechnology (Hussain et al., 2016). The biosynthesis method of nanoparticle synthesis, in addition to reduced toxicity has several advantages like being eco-friendly, cost-effective, and simplicity and compatibility (Hariharan et al., 2017). There are many ways of green biosynthesis of nanoparticles including plant extracts, bacteria, fungi, enzymes and seaweeds (Varghese et al., 2020). Biosynthesis methods of TiO_2 nanoparticle production especially using plant extracts are of major importance due to major reasons like nontoxicity and environmental friendliness (Bekele et al., 2020).

Among the biosynthesis methods, plant extract-based biosynthesis is a major method for the production of nanoparticles as it results in higher yield and faster process. Other advantages of plant-based biosynthesis include, plants are cheaper than enzymes and chemicals, they can withstand various environmental abiotic stress, don't require any special incubation methods and are eco-friendly. In-plant extract based biosynthesis, the metal ions in precursor-like TiO_4, $TiO(OH)_{2\ are}$ reduced to form nucleation centres and it sequesters additional metal ions resulting in the formation of TiO_2 nanoparticles (Varghese et al., 2020).

The microbial metabolites produced are major agents in capping, bioreducing, and stabilizing properties that are capable of improving the performance of nanoparticle synthesis (de Dicastillo et al., 2020). Biosynthesis of TiO_2 NPS play a major role in the application of nanotechnology to a variety of bacterial pathogens including *E. coli, K. pneumoniae, P. vulgaris,* MRSA and *S. aureus.* Thus, TiO_2 NPs that are synthesized by biological method including plant extract, bacteria, fungi and enzymes are of major focus in medicine, clinical applications and combating drug-resistant pathogens (Hussain et al., 2016).

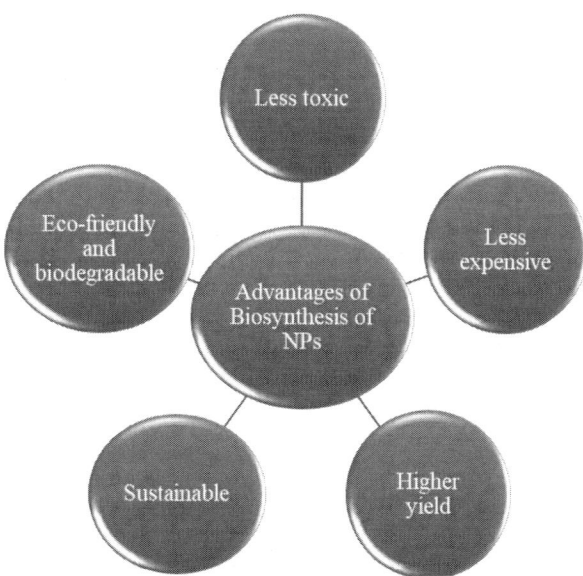

Figure 1. Advantages of biosynthesis of Nanoparticles over chemical method of synthesis.

METHODS OF CHARACTERIZATION OF TiO_2 NANOPARTICLES

After the synthesis of TiO_2 NPs, it must be characterized for studying its structure, size and other characteristics. There are various methods employed for the characterization of synthesized nanoparticles. They can be classified into three major methods namely spectroscopic techniques, microscopic techniques and diffraction techniques (Khare et al., 2019).

1. *Spectroscopic techniques:* These methods are used to examine size and shape of the NPs. As TiO_2 NPs show surface plasmon resonance, its absorption and emission spectra can be measured using UV-Visible spectrophotometry and its association with the formation of spherical nanoparticles can be studied. Other spectroscopic techniques like infrared spectroscopy and Fourier Transform Infrared spectroscopy can also be used. FT-IR is the most commonly used spectroscopic method for determining the structural groups on the surface of NPs (Khare et al., 2019).
2. *Microscopic techniques:* Microscopic techniques are mainly used to study the morphology, structure and size of the synthesized NPs. Electron microscopy is the most employed method that is very efficient in analysis of NPs as it provides magnification of the specimen to a large extent which enables visualization of NPs in the nanometre range. Scanning Electron Microscopy and Transmission Electron Microscopy can be used for NP characterization (Hussain et al., 2016; Khare et al., 2019).
3. *Diffraction techniques:* Diffraction methods like Dynamic light scattering, X-ray diffraction (XRD) and zeta potential measurement can be used to provide information about the crystal structure of the NPs. XRD gives information about translational symmetry, size and phase identification of metallic and metal oxide NPs (Hussain et al., 2016).

ANTIBACTERIAL ACTIVITY OF BIOSYNTHESIZED TiO$_2$ NANOPARTICLES

TiO$_2$ NPs are one of the major types of nanoparticles that are being studied for their antibacterial activity against a wide range of bacteria including gram positive and gram negative bacteria, drug-resistant bacteria and even extremely drug-resistant bacteria. In comparison with chemically synthesized TiO$_2$ NPs, biosynthesized TiO$_2$ NPs using plant extracts, microbes and their enzymes have exhibited higher antibacterial activity. In addition to various advantages of the green method of TiO$_2$ NPs synthesis like less toxicity and being eco-friendly, it is noticeable as it exhibits higher antibacterial activity than the commercial antibacterial and chemically synthesized TiO$_2$ NPs. Biosynthesis of TiO$_2$ NPs can be mediated by extracts of various parts of the plant-like root, leaf, flower, seed and can also be mediated by microbes like bacteria, fungi and their enzymes.

PLANT MEDIATED BIOSYNTHESIS OF TiO$_2$ NANOPARTICLES AND THEIR ANTIBACTERIAL ACTIVITY

Various plant part extracts like complete plant extract, leaf, root, fruit peel, twig and bud extracts can be used to mediate the biosynthesis of TiO$_2$ NPs from different precursors that exhibit significant antibacterial activity against a wide range of pathogens. The antibacterial activity can be studied by various methods like disc diffusion, Minimum Inhibition Concentration and Minimum Bactericidal concentration. A brief account of plant-mediated biosynthesis of TiO$_2$ NPs and their antibacterial activity is given in Table 1.

TiO$_2$ nanoparticles were biosynthesized using plant extract of *Hypheae Thiebeace* and *Anannos Seneglensis* by the activity of phytochemical compounds like flavonoids, steroids, terpenoids and polyphenols. Antibacterial activity of the synthesized TiO$_2$ nanoparticles was measured in terms of minimum bactericidal concentration and minimum inhibitory concentration by well diffusion and dilution method respectively. It exhibited significant antibacterial activity against *Bacillus subtilis, Staphylococcus aureus, Escherichia coli* and *Salmonella typhi*. Biosynthesized TiO_2 NPs showed significant antibacterial activity than commercial TiO_2 NPs and commercial antibiotics like ampicillin, tobramycin and erythromycin (S. E. D. Hassan et al., 2018).

Biosynthesized TiO_2 NPS using *Trigonella foenum-graecum* extract was analysed for their antibacterial activity against major pathogens like *Staphylococcus aureus, Enterococcus faecalis, Klebsiella pneumoniae, Streptococcus faecalis, Pseudomonas aeruginosa, Escherichia coli, Proteus vulgaris, Bacillus subtilis* and *Yersinia enterocolitica*. Biosynthesized spherical TiO_2 NPs of size 20-90nm exhibited significant antibacterial activity against *E. faecalis, S. aureus, S. faecalis* with a zone of inhibition of 11.4, 11.2 and 11.6 mm respectively and antibacterial activity was observed against all the tested bacteria. Dissolving outer membrane due to the hydroxyl group presence, cell wall structure difference between gram-positive and gram-negative bacteria were found to be few mechanisms associated with antibacterial activity (Subhapriya & Gomathipriya, 2018).

TiO_2 NPS when treated with *Garcinia zeylanica* extract, it was observed that the antibacterial activity against Methicillin-resistant *S. aureus* was enhanced in comparison to TiO_2 and plant extract individually. Antibacterial activity was measured by spread plate method and analysis of colony-forming units. After 24 hours of incubation in presence of sunlight, CFU of MRSA was significantly reduced due to the antibacterial effect of *Garcinia zeylanica* extract effect on TiO_2 NPs. Phytochemical analysis of the plant extract revealed the presence of tannins, cardiac glycoside, saponins, carbohydrates and coumarin which may be responsible for the antibacterial activity (Senarathna et al., 2017).

Biosynthesis of TiO_2 NPs using aqueous extract of *Psidium guajava* was carried out and its antibacterial activity against 5 different pathogens was tested by disc diffusion assay. The highest antibacterial activity was observed against *S. aureus* and *E. coli* with zones of inhibition of 25mm and 23 mm respectively. This biosynthesized TiO$_2$ reported higher antibacterial activity than the commercial antibiotic tetracycline paving a new way for treating antimicrobial resistant *S. aureus* (Santhoshkumar et al., 2014). TiO$_2$ NPs biosynthesized using an extract of *Vitex negundo Linn* reported antibacterial activity against *S. aureus* and *E. coli* (Ambika & Sundrarajan, 2016).

TiO$_2$ NPs synthesized from *Trichoderma citrinoviride* extract were tested for their significant antibacterial activity against extremely drug-resistant *Pseudomonas aeruginosa* (Arya et al., 2021). *Cola nitida* extract was used for the production of TiO$_2$ NPs and it was found to exhibit significant antibacterial activity against *E. coli* and *K. pneumonia*. Significant dye degradation activity, anti-coagulant activity and antioxidant activities were also reported (Akinola et al., 2020). *Cynodon dactylon* based biosynthesis of TiO$_2$ NPs resulted in spherical nanoparticles of size 16nm and was found effective against *E. coli*. *Diospyros ebenum* based spherical TiO$_2$ NPs of size 10-12nm were also effective against *E. coli* (Hariharan et al., 2017).

TiO$_2$ NP's antibacterial activity can be enhanced by the use of plant extracts like *Bauhinia variegata and Tinospora cordifolia*. The nanocomposite of the aforesaid plant extract (various solvents) with TiO$_2$ NPs was tested for their significant antibacterial activity against *E. coli* and *E. faecalis* by disc diffusion method. Plant extract TiO$_2$ NPs nanocomposites showed significant antibacterial activity against both the tested pathogens in comparison with TiO$_2$ NPs and plant extract. The highest zone of inhibition was exhibited by aqueous extract of *Bauhinia variegata*/TiO$_2$ nanocomposite against *E. faecalis* and benzene extract of *Tinospora cordifolia* against *E. coli* (Maurya et al., 2012).

Silver doped TiO$_2$ NPs were synthesized using the aqueous extract of *Acacia nilotica* and its antibacterial activity against a variety of pathogenic bacteria was tested using well diffusion method. Spherical Ag/TiO$_2$ NPs exhibited significant antibacterial activity against *E. coli* followed by Methicillin-resistant *S. aureus* and *P. aeruginosa*. Glutathione and lipid peroxidation levels in cells were analyzed to confirm the antibacterial mechanism of the Ag/TiO$_2$ NPs. Glutathione level was reduced in the treated bacterial cells which increased ROS generation and lipid peroxidation eventually leading to cell death (Rao et al., 2019).

Leaf Extract Mediated TiO$_2$ Nanoparticles

TiO$_2$ synthesized from leaf extract of *Azadirachta indica* exhibited significant antibacterial activity against a variety of pathogens like *Escherichia coli, Bacillus subtilis, Salmonella typhi* and *Klebsiella pneumoniae*. Better antibacterial activity was produced by TiO$_2$ NPs than the TiO$_2$ compound. FTIR analysis reported the presence of terpenoids, flavonoids and proteins which were associated with synthesis of TiO$_2$ NPs using leaf extract. TiO$_2$ that exhibited antibacterial activity was observed to be spherical on Electron microscopy analysis. The lowest MIC of 10.42 µg/mL and MBC of 83.3 µg/mL was observed against *Salmonella typhi, E. coli* and *K. pneumoniae* respectively (Thakur et al., 2019).

Pithecellobium dulce and *Lagenaria siceraria* leaf extracts were used for biosynthesis of TiO$_2$ NPs and its antibacterial activity was tested against gram-positive and gram negative bacteria like *Streptococcus pyogenes, Bacillus subtilis, Pseudomonas aeruginosa, E. coli* and *Staphylococcus aureus*. TiO$_2$ NPs biosynthesized from both the leaf extracts showed significant antibacterial activity against all the tested bacteria with the highest activity against *B. subtilis* and zones were higher than commercial TiO$_2$ NPs. *P. dulce* based TiO$_2$ NPs exhibited higher antibacterial activity than the TiO$_2$ NPs synthesized using *L. siceraria*. A similar pattern of antibacterial activity was observed against other bacteria tested (Kalyanasundaram & Prakash, 2015).

Biosynthesis of TiO$_2$ NPs using *Glycosmis cochinchinensis* leaf extract was carried out and its antibacterial activity was tested against both gram-positive and gram-negative organisms including *Staphylococcus saprophyticus, Bacillus subtilis, E. coli* and *P. aeruginosa*. TiO$_2$ NPs synthesized using leaf extract showed significantly higher antibacterial activity than chemically synthesized TiO_2 NPS and leaf extract alone. The highest antibacterial activity was reported against *P. aeruginosa* by the disc diffusion method. As reported in other studies, higher antibacterial activity was observed in gram-negative bacteria due to their cell wall structure (Rosi & Kalyanasundaram, 2018).

In a study, TiO_2 NPs synthesized using leaf extract of *Cynodon dactylon* were tested for their antibacterial activity against *E. coli* using a good diffusion method and significant antibacterial activity was observed in comparison to positive control gentamycin (Hariharan et al., 2017). Zinc oxide and TiO_2 NPs were biosynthesized using *Ficus religiosa* leaf extract (aqueous). Its antibacterial activity was tested against *E. coli* and *S. aureus*. Significant antibacterial activity was observed in both nanoparticles (Soni & Dhiman, 2020).

Mentha arvensi leaf extract-based biosynthesis of TiO_2 NPs yielded spherical nanoparticles in size of 20-70nm. Its antibacterial activity was analyzed using disc diffusion assay and significant activity was observed against *Proteus vulgaris* but less than the commercial antibiotics used as a positive control. Further purification of bioactive compounds from the leaf extract can be used for TiO_2 biosynthesis which may enhance the antibacterial activity (W. Ahmad et al., 2020). Similarly, TiO2 NPs synthesized using leaf extract of *Avicennia marina* showed significant antibacterial activity against aquatic pathogens (Shahin Lefteh et al., 2020).

Parthenium hysterophorus leaf aqueous extract was used for the biosynthesis of TiO_2 NPs using TiO_4. Biologically active compounds like phenols, fluoroalkanes, alcohols and alkanes and were involved in the bio-reduction of TiO_4 into TiO_2. Antimicrobial efficacy of TiO_2 NPs has been tested on clinical pathogens *Staphylococcus aureus, Escherichia coli, Pseudomonas aeruginosa, Proteus vulgaris, Klebsiella pneumoniae* and *Staphylococcus epidermidis*. The highest antibacterial activity was exhibited against *Staphylococcus epidermidis, Pseudomonas aeruginosa* and *Proteus vulgaris* with a zone of inhibition 21, 17.6 and 15.6mm respectively (Thandapani et al., 2018). *Morinda citrifolia* leaf extract was used for the production of quasi-spherical shaped TiO_2 NPs using $TiCl_4$ as a precursor. It exhibited significant antibacterial activity against *S. aureus, B. subtilis, E. coli, P. aeruginosa* tested by the agar diffusion method (Sundrarajan et al., 2017).

Fruit Peel Extract Mediated TiO_2 Nanoparticles

Peel extract of *Citrus aurantium* fruit was used as a reducing and capping agent for the biosynthesis of TiO_2 NPs and it was characterized by various methods like XRD, FT-IR, EDAX, TEM and AFM. Its antibacterial activity was tested against both gram-positive and gram-negative bacteria like *S. aureus, Staphylococcus epidermidis, P. aeruginosa* and *K. pneumonia*. The highest zone of inhibition was observed for *S. aureus* followed by *S. epidermidis, P. aeruginosa* and *K. pneumoniae*. Major mechanisms proposed for antibacterial activity of biosynthesized TiO_2 NPs include dissolving bacterial outer membrane due to the presence of hydroxyl groups resulting in bacterial cell death. High antibacterial activity was observed against gram-negative bacteria than the gram-positive bacteria which may be due to the difference in the peptidoglycan layer in the cell wall. A thin layer of peptidoglycan layer in gram-negative bacteria makes it easier for metal to get in contact with bacterial cells easier than the gram-positive cells having thicker peptidoglycan cell wall Similarly, the electrostatic force between the negative bacterial cell and positive NPs and also the deactivation of cellular enzymes leading to the formation of pits in bacterial cell wall also leads to cell death. (Punitha et al., 2020).

Prunus Persia L. (Peach), *Prunus domestica L. (*Plum*)* and *Actinidia deliciosa* (Kiwi)) peel extract was used for biosynthesis of TiO_2 NPS and its antibacterial activity was tested against pathogens like *Escherichia coli, Staphylococcus aureus, Pseudomonas aeruginosa* and *Bacillus substilis* by disc diffusion method. Significant antibacterial activity was exhibited by all three biosynthesized TiO_2 NPs. The highest antibacterial activity was reported against *E. coli* with a zone of inhibition range of

4 -20 mm, 3 -18 mm and 5-20 mm for plum, peach and kiwi peel extract-based TiO_2 NPs (Ajmal et al., 2019).

The nanocomposite of pristine pomegranate peel extract (PPP) with TiO_2 exhibited significant antibacterial activity against *P. aeruginosa, E. coli and S. aureus*. Significant antibacterial activity was reported on various assays like minimum inhibition concentration (MIC), Minimum bactericidal concentration (MBC), well diffusion method and Fluorescence microscopic study for live/dead cell viability (Abu-Dalo et al., 2019).

TiO_2 NPs were also synthesized using orange fruit peel by the green biosynthesis method. Its antibacterial activity was tested against *Escherichia coli, Staphylococcus aureus* and *Pseudomonuas aeruginosa* and significant antibacterial activity were observed. Green synthesized TiO_2 exhibited higher antibacterial activity than chemically synthesized nanoparticles. Out of the pathogens tested in this study, *Pseudomonas aeruginosa* possessed the highest zone of inhibition. Similar to other studies, antibacterial activity was attributed to the production of ROS and pore formation in the cell wall (Amanulla & Sundaram, 2019).

Root Extract Mediated TiO_2 Nanoparticles

TiO_2 NPs were biosynthesized using root extract of *Kniphofia foliosa* using titanium tetrabutoxide as a precursor. Three different ratios of titanium tetrabutoxide to the root extract including 1:2, 1:1 and 2:1 was used for the synthesis of TiO_2 NPs. Antibacterial activity against human pathogen bacteria strains of *Staphylococcus aureus, Escherichia coli, Klebsiella pneumonia* and *Streptococcus pyogenes* by disc diffusion method. Among these ratios, the 1:1 ratio of titanium tetrabutoxide to the root extract synthesized TiO_2 exhibited significant antibacterial activity against both gram-positive and gram-negative pathogens. 1:1 ratio exhibited better antibacterial activity than the other two due to the smaller average crystalline size, spherical size of TiO_2 NPS which was identified based on SEM analysis. The composition of Titanium tetrabutoxide to root extract also played a significant role in deciding the antibacterial efficacy of the TiO$_2$ NPs synthesized. Few mechanisms were proposed for the antibacterial activity of the biosynthesized TiO$_2$ NPS. Bacterial cell membrane breakage was caused due to the reactive oxygen species production due to the generation of O^{2-} and superoxide radicals via the transfer of electrons by biomolecules in root extract. Similarly, metal oxides and bacteria carry positive and negative charges respectively which causes electrostatic attraction between them resulting in oxidation and death of bacteria. TiO$_2$ are also involved in other antibacterial mechanisms like deactivating cellular enzymes, destruction of the outer membrane of bacterial cells which causes leakage of intracellular components leading to cell death (Bekele et al., 2020).

TiO$_2$ NPs were synthesized using root extracts of *Glycyrrhiza glabra* by green synthesis method using titanium oxysulfate precursor. 60-140 nm size ranged TiO$_2$ NPs synthesized were tested for their antibacterial activity against *K. pneumonia* and *S. aureus*. TiO$_2$ NPs synthesized exhibited significant activity against *K. pneumonia*. Further in vivo toxicity study was carried out using an embryonic model of zebrafish and biocompatibility was observed without any malformation after treatment with TiO$_2$ NPs (Bavanilatha et al., 2019).

Flower Extract Mediated TiO$_2$ Nanoparticles

Jasmine flower extract was used for the biosynthesis of TiO$_2$ NPs and 32-48nm-sized spherical nanoparticles were formed. Alkaloids, coumarins and flavonoids in the jasmine flower were associated with the reduction of titanium tetra isopropoxide into TiO$_2$ NPs. Its antibacterial activity was studied in comparison to chemically synthesized TiO$_2$ and it was tested *against Escherichia coli, Staphylococcus aureus* and *Klebsiella pneumoniae*. Green synthesized nanoparticles showed a higher zone of inhibition than chemically synthesized nanoparticles. Higher antibacterial activity was exhibited against *E. coli and K. pneumonia* than *S. aureus*. Various mechanisms like ROS production, cell wall degradation. The m TiO_2 NPs reported a potent antibacterial activity (Aravind et al., 2021). Biosynthesis of TiO_2 NPs was carried out using flower extract of hibiscus using Titanium Oxysulfate as a precursor. Biosynthesized TiO_2 NPs were tested for their antibacterial activity against *Vibrio cholerae, Pseudomonas aeruginosa* and *Staphylococcus aureus* using disc diffusion assay and in comparison with chemically synthesized TiO_2 NPs. Biosynthesized NPs exhibited higher antibacterial activity with a higher zone of inhibition in comparison to chemically synthesized TiO_2 NPs (Kumar et al., 2014).

Other Plant Part Extracts Mediated TiO_2 Nanoparticles

TiO_2 NPs were green synthesized using the twig extract of *Azadirachta indica, Ficus benghalensis* and bud extract of *Syzygium aromaticum*. Antibacterial activity of the synthesized TiO_2 NPs against *Streptococcus mutans* and *Citrobacter freundii* was measured in terms of MIC using the Resazurin dye method and disc diffusion method. Significant antibacterial activity was exhibited by synthesized TiO_2 than the crude extract of the plants used. Higher antibacterial activity was observed against *C. freundii* than *S. mutans*. The potential antimicrobial mechanism was proposed to be due to the small size and large surface area of the TiO_2 NPs which releases Ti^{4+} ions from TiO_2. These ions increase oxidative stress in the bacteria causing rupturing of the cell wall and ultimately leading to cell death (Achudhan et al., 2020).

TiO_2 NPs synthesized using *Vigna unguiculata* seed extract and were characterized by FTIR and SEM. Its potential antibacterial activity was tested against a wide range of gram-positive and gram-negative bacteria by a good diffusion method. Out of twelve pathogens tested, significant antibacterial activity was observed against *E. coli, Salmonella, Enterobacter* and *Serratia marcescens* with the highest activity against *S. marcescens*. Similarly, its anticancer activity against osteosarcoma cells was also observed (Chatterjee et al., 2017).

Table 1. Antibacterial activity of Plant mediated biosynthesized TiO$_2$ Nanoparticles

Source	Plant part	Precursor	Target bacteria	Method	Reference
Kniphofia foliosa	Root	Titanium Tetrabutoxide	*Staphylococcus aureus, Escherichia coli, Klebsiella pneumonia and Streptococcus pyogenes*	disc diffusion method	(Bekele et al., 2020)
Hypheae thiebeace and *Ananmos seneglensis*	Plant extract	TiCl$_4$	*Bacillus subtilis, S. aureus, E.coli, Salmonella typhi.*	Well diffusion and dilution method	(H. Hassan et al., 2019)
Azadirachta indica	Leaf	Titanium dioxide	*E. coli, B. subtilis, S. typhi* and *K. pneumoniae*	MIC, MBC, Electron microscopy	(Thakur et al., 2019)
Pomegranate	Peel	Titanium dioxide	*P. aeruginosa, E. coli and S. aureus*	MIC, MBC, Fluorescence microscopy	(Abu-Dalo et al., 2019)
Hibiscus	Flower	Titanium Oxysulfate	*Vibrio cholerae, P. aeruginosa* and *S. aureus*	Disc diffusion	(Kumar et al., 2014)
Azadirachta indica, Ficus benghalensis	Twig	Titanium isopropoxide	*Streptococcus mutans* and *Citrobacter freundii*	MIC using Resazurin dye method and disc diffusion	(Achudhan et al., 2020)
Syzygium aromaticum	Bud	Titanium isopropoxide	*S. mutans and C. freundii*	MIC using Resazurin dye method and disc diffusion	(Achudhan et al., 2020)

MICROBES AND THEIR ENZYMES MEDIATED TiO_2 NANOPARTICLES BIOSYNTHESIS AND THEIR ANTIBACTERIAL ACTIVITY

Even though plant extracts are the major source studied for the biosynthesis of TiO_2 NPs, microbial sources like bacteria, fungi and their enzymes have also been studied for their potential to synthesize TiO_2 NPs from their precursor and a brief description is provided in Table 2.

Table 2. Microbes mediated biosynthesized TiO2 Nanoparticles and their antibacterial activity

Microbial Source	TiO_2 Nanoparticles size (nm)	Target organism	Reference
Aeromonas hydrophila	40.5	A. hydrophila, Pseudomonas aeruginosa, Enterococcus faecalis, Staphylococcus aureus, Escherichia coli and Streptococcus pyogenes.	(Jayaseelan et al., 2013)
Aspergillus flavus	62–74	E.coli, S.aureus	(Rajakumar et al., 2012)
Bacillus mycoides	40-60	E.coli	(Órdenes-Aenishanslins et al., 2014)
Bacillus subtilis	10-30	Aquatic biofilm	(Dhandapani et al., 2012)
Lactobacillus crispatus	70.98	K. pneumonia, Acinetobacter baumannii, S. aureus, E. coli and Morganella morganii.	(Ibrahem et al., 2014)

In addition to plant extracts, the fungus can also be used as a reducing and capping agent for the biosynthesis of TiO_2 NPS. *Aspergillus flavus* MTCC culture was used for the production of TiO_2 NPs. The fungal biomass for the biosynthesis of TiO_2 NPs was achieved by collection and filtration of mycelium and adding it to the TiO_2. It resulted in the formation of TiO_2 NPs. On analysis of its antibacterial activity by minimum inhibitory concentration (MIC) and well diffusion assay, the highest antibacterial activity was observed against *E. coli* in both MIC and well diffusion assay at MIC of 40 μg ml−1) and zone of inhibition of 35mm followed by *S. aureus*. The presence of various quinones in *Aspergillus* was proposed as the major driving factor in fungus-mediated production of TiO_2 NPs (Rajakumar et al., 2012).

Several studies have been reported for the biosynthesis of TiO_2 NPs using bacteria as a reducing agent. *Bacillus subtilis* was used for biosynthesis of TiO_2 NPs which yielded spherical nanoparticles of size 30-40 nm. It exhibited significant antibacterial activity against *E. coli* which was confirmed by a reduction in colony-forming units on the increased concentration of TiO_2 NPs. Growth curve analysis also proved the antibacterial activity of the biosynthesized TiO_2 NPs. A potential mechanism for antibacterial activity was proposed as the inactivation of cellular enzymes and DNA due to binding to electron-donating groups like carboxylates, amides, thiols, etc. This binding caused small pores in the cell walls of the bacteria which causes cellular content to flow out as a result of a change in cellular permeability and eventually leading to death (Singh, 2016).

Similarly, another novel, cost-effective and reproducible bacteria *Aeromonas hydrophila* was used as a reducing and capping agent for the biosynthesis of TiO_2 NPs and smooth, spherical and uneven TiO_2 NPs of size 40.5nm was synthesized. GC-MS analysis of the bacteria revealed the presence of major compounds like uric acid, glycyl-L-glutamic acid, glycyl-L-proline and l-Leucyl-d-leucine. Out of these 4 major compounds, studies have reported that glycyl-L-proline may act as a potential capping agent in the synthesis of TiO_2 NPs. Biosynthesized TiO_2 NPs were tested for their antibacterial activity against *A. hydrophila, Pseudomonas aeruginosa, Streptococcus pyogenes, Escherichia coli, Staphylococcus aureus* and *Enterococcus faecalis*. The highest antibacterial activity was exhibited against *S. aureus* and *S. pyogenes* in the good diffusion method (Jayaseelan et al., 2013).

Planomicrobium sp., a gram-positive bacteria was used to synthesize TiO_2 NPs and its antibacterial activity against *Bacillus subtilis* and *Klebsiella planticola* and significant antibacterial activity was observed.

Higher activity was observed against *B. subtilis* (Rajeshkumar, n.d.). TiO_2 NPs synthesized using *Bacillus mycoides,* an environmental isolate yielded 40 – 60nm size nanoparticles from precursor $TiO(OH)_2$ and it showed significant antibacterial activity against *E.coli* (Órdenes-Aenishanslins et al., 2014). *Bacillus subtilis* was used as a reducing agent for the synthesis of TiO_2 NPs (size 10-30nm) from potassium hexafluorotitanate exhibited significant suppression against the growth of aquatic biofilm (Dhandapani et al., 2012).

Similarly, *S. aureus* based synthesis of TiO_2 NPs exhibited significant antibacterial activity at 10-15 mg/ml concentration (Landage et al., 2020). *Lactobacillus crispatus* based TiO_2 NPs of size 70.98 nm showed significant antibacterial activity against *K. pneumonia, S. aureus, Acinetobacter baumannii, E. coli* and *Morganella morganii* (Ibrahem et al., 2014).

The alpha-amylase enzyme was also used as a potential reducing and capping agent for the synthesis of TiO_2 NPs from $TiO(OH)_2$ precursors. Synthesized TiO_2 NPs were tested for their antibacterial activity against *S. aureus* and *E. coli* by agar diffusion assay, growth curve study and its effect on the bacterial cell was analyzed using TEM and confocal microscopy analysis. Significant antibacterial activity was observed with 62.50 μg/ml for both gram-positive and gram-negative bacteria was observed. Further TiO_2 has exhibited the antibacterial effect by damaging the cell wall of bacteria due to the production of ROS and the formation of pores. It was evidenced by analysis with confocal microscopy and TEM analysis (R. Ahmad et al., 2015).

In addition to pure TiO_2 NPs, various metals and their compounds can be doped with TiO_2 NPs to enhance their antibacterial activity against a wide range of pathogens. Fe_3O_4 -TiO_2 magnetic nanoparticles reduced the survival rate of Methicillin-resistant *S. aureus* from 82.4% to 7.13%. Similarly high reduction in survival ratio of *Staphylococcus saprophyticus, Streptococcus pyogenes* was exhibited by Fe_3O_4 -TiO_2 magnetic nanoparticles. TiO_2 nanotubes of size 20nm reduced the growth of *E. coli* and *S. aureus* by 97 and 99% respectively. Similarly, copper doped TiO_2 NPs of size 20nm inhibited the growth of *Mycobacterium smegmatis* and *Shewanella oneidensis MR-1* by 47% and 11% respectively (de Dicastillo et al., 2020).

ANTIBACTERIAL MECHANISM OF TiO$_2$ NANOPARTICLES

There are various target organelles that TiO$_2$ NPs can attack in bacteria to inhibit its growth and cell death. One of the major mechanisms that are proposed in most of the studies is the production of Reactive Oxygen Species (ROS) which causes cell wall damage leading to cell death. Other than that, any mechanisms can be observed for its antibacterial activity (de Dicastillo et al., 2020).

Cell Wall

A major factor to be noted in the cell wall as the target of TiO$_2$ NPs is the difference in structure between gram-positive and gram-negative bacteria. As gram-negative bacteria have a thin peptidoglycan layer compared to gram-positive bacteria, there can be variation in sensitivity to the TiO$_2$ NPs. The cell wall of *E. coli* is reported to be sensitive to peroxidation caused by TiO$_2$ NPs. Sometimes TiO$_2$ NPs can also cause genetic issues by causing lower expression of genes encoding for proteins involved in the metabolism of lipopolysaccharide and peptidoglycan, pilus biosynthesis and transmembrane proteins (Kubacka et al., 2014).

Cell Membrane

As already stated, the difference in cell wall structure between gram-positive and gram-negative bacteria plays a major role in damage caused by TiO$_2$ NPs in the cell membrane. Major damage in the cell membrane is caused due to the oxidation of phospholipids due to the production of ROS such as hydroxyl radicals and hydrogen peroxide which results in a change of permeability of cell membrane, increase in membrane fluidity, cellular content leakage, and eventually leading to cell death (Khezerlou et al., 2018; Pavlović et al., 2019).

Effect in the Respiratory Chain

Mitochondria produces enzymes like superoxide dismutase to compensate for normal levels of ROS produced in the cell. In the case of ROS production in cells due to TiO_2 NPs, this enzyme cannot cope up to attenuate the damage and the respiratory chain will be inhibited. Also as a genetic issue of TiO_2 NPs, genes that are associated with energy production in mitochondria and oxygen uptake pathways (Kubacka et al., 2014).

DNA

DNA is also a major part where ROS causes damage by causing mutation and DNA breaks by Fenton reaction etc. Mitochondrial DNA is more easily damaged due to ROS production than nuclear DNA. DNA damage repair mechanisms are upregulated in case of exposure to TiO_2 NPs which indicated that DNA damage is also a major antibacterial mechanism exhibited by TiO_2 NPs (de Dicastillo et al., 2020).

Deficiency in Iron and Inorganic Phosphate (Pi)

Iron is one of the major nutrients required for maintaining homeostasis in the bacterial cell. Iron concentration in bacterial cells is regulated by various mechanisms like siderophores. On treatment with TiO_2 NPs to *Pseudomonas brassicacearum,* ICP MS analysis revealed lower expression of genes related to siderophore production and iron transport proteins (Liu et al., 2016). Similarly, inorganic phosphate is also required to maintain normal cell functioning. One of the major regulons involved in Pi regulation in a bacterial cell is Pho regulon which was significantly expressed lower after TiO_2 NPs treatment (Haddad et al., 2009).

Cell to Cell Communication

TiO_2 NPs can also affect cell to cell communication by directly oxidizing components of cell signaling pathways and can change the gene expression by interfering with transcription factors. TiO_2 NPs also affect quorum sensing signaling molecules related to pathogenesis and biofilm production and can bacterial growth inhibition. This was evidenced on SEM analysis of TiO_2 NPs treated *P. aeruginosa* (Kubacka et al., 2007).

CONCLUSION

The biosynthesis of titanium dioxide nanoparticles is a major area of research. This can pave way for the huge interest in the field of pharmaceuticals especially antibacterial drugs. Biosynthesis can be mediated by various plant extracts, bacteria, fungi and their enzymes. As antibiotic resistance is emerging as a threat, metal oxide-based nanoparticles especially TiO_2 NPs provide a novel antibacterial therapy alternative. They are much preferred due to their photocatalyst activity, less toxicity and it is Generally Recognized as a Safe (GRAS) substance (Gold et al., 2018). They provide a novel solution for a long-term, effective antibacterial and antibiofilm agent by the major mechanism of production of ROS, membrane instability and DNA damage. In the future, TiO_2 NPs are predicted to be combined with antibacterial for optimal activity due to their antibacterial nature.

REFERENCES

Abu-Dalo, M., Jaradat, A., Albiss, B. A., & Al-Rawashdeh, N. A. (2019). Green synthesis of TiO_2 NPs/pristine pomegranate peel extract nanocomposite and its antimicrobial activity for water disinfection. *Journal of Environmental Chemical Engineering*, *7*(5), 103370.

Achudhan, D., Vijayakumar, S., Malaikozhundan, B., Divya, M., Jothirajan, M., Subbian, K., González-Sánchez, Z. I., Mahboob, S., Al-Ghanim, K. A., & Vaseeharan, B. (2020). The antibacterial, antibiofilm, antifogging and mosquitocidal activities of titanium dioxide (TiO_2) nanoparticles green-synthesized using multiple plants extracts. *Journal of Environmental Chemical Engineering*, *8*(6), 104521.

Ahmad, R., Mohsin, M., Ahmad, T., & Sardar, M. (2015). Alpha amylase assisted synthesis of TiO_2 nanoparticles: Structural characterization and application as antibacterial agents. *Journal of Hazardous Materials*, *283*, 171–177.

Ahmad, W., Jaiswal, K. K., & Soni, S. (2020). Green synthesis of titanium dioxide (TiO_2) nanoparticles by using *Mentha arvensis* leaves extract and its antimicrobial properties. *Inorganic and Nano-Metal Chemistry*, *50*(10), 1032–1038.

Ajmal, N., Saraswat, K., Bakht, M. A., Riadi, Y., Ahsan, M. J., & Noushad, M. (2019). Cost-effective and eco-friendly synthesis of titanium dioxide (TiO_2) nanoparticles using fruit's peel agro-waste extracts: Characterization, in vitro antibacterial, antioxidant activities. *Green Chemistry Letters and Reviews*, *12*(3), 244–254.

Akinola, P. O., Lateef, A., Asafa, T. B., Beukes, L. S., Hakeem, A. S., & Irshad, H. M. (2020). Multifunctional titanium dioxide nanoparticles biofabricated via phytosynthetic route using extracts of *Cola nitida*: Antimicrobial, dye degradation, antioxidant and anticoagulant activities. *Heliyon*, *6*(8), e04610.

Amanulla, A. M., & Sundaram, R. (2019). Green synthesis of TiO_2 nanoparticles using orange peel extract for antibacterial, cytotoxicity and humidity sensor applications. *Materials Today: Proceedings*, *8*, 323–331.

Ambika, S., & Sundrarajan, M. (2016). [EMIM] BF4 ionic liquid-mediated synthesis of TiO$_2$ nanoparticles using *Vitex negundo Linn* extract and its antibacterial activity. *Journal of Molecular Liquids*, *221*, 986–992.

Aravind, M., Amalanathan, M., & Mary, M. S. M. (2021). Synthesis of TiO$_2$ nanoparticles by chemical and green synthesis methods and their multifaceted properties. *SN Applied Sciences*, *3*(4), 1–10.

Arya, S., Sonawane, H., Math, S., Tambade, P., Chaskar, M., & Shinde, D. (2021). Biogenic titanium nanoparticles (TiO 2 NPs) from *Tricoderma citrinoviride* extract: Synthesis, characterization and antibacterial activity against extremely drug-resistant *Pseudomonas aeruginosa*. *International Nano Letters*, *11*(1), 35–42.

Bavanilatha, M., Yoshitha, L., Nivedhitha, S., & Sahithya, S. (2019). Bioactive studies of TiO$_2$ nanoparticles synthesized using *Glycyrrhiza glabra*. *Biocatalysis and Agricultural Biotechnology*, *19*, 101131.

Bekele, E. T., Gonfa, B. A., Zelekew, O. A., Belay, H. H., & Sabir, F. K. (2020). Synthesis of titanium oxide nanoparticles using root extract of K*niphofia foliosa* as a template, characterization, and its application on drug resistance bacteria. *Journal of Nanomaterials*, *2020*.

Chatterjee, A., Ajantha, M., Talekar, A., Revathy, N., & Abraham, J. (2017). Biosynthesis, antimicrobial and cytotoxic effects of titanium dioxide nanoparticles using *Vigna unguiculata* seeds. *International Journal of Pharmacognosy and Phytochemical Research*, *9*(1), 95–99.

de Dicastillo, C. L., Correa, M. G., Martínez, F. B., Streitt, C., & Galotto, M. J. (2020). Antimicrobial effect of titanium dioxide nanoparticles. In *Antimicrobial Resistance-A One Health Perspective*. IntechOpen.

Dhandapani, P., Maruthamuthu, S., & Rajagopal, G. (2012). Bio-mediated synthesis of TiO$_2$ nanoparticles and its photocatalytic effect on aquatic biofilm. *Journal of Photochemistry and Photobiology B: Biology*, *110*, 43–49.

Gold, K., Slay, B., Knackstedt, M., & Gaharwar, A. K. (2018). Antimicrobial activity of metal and metal-oxide based nanoparticles. *Advanced Therapeutics*, *1*(3), 1700033.

Haddad, A., Jensen, V., Becker, T., & Häussler, S. (2009). The Pho regulon influences biofilm formation and type three secretion in *Pseudomonas aeruginosa*. *Environmental Microbiology Reports*, *1*(6), 488–494.

Hariharan, D., Srinivasan, K., & Nehru, L. C. (2017). Synthesis and characterization of TiO$_2$ nanoparticles using cynodon dactylon leaf extract for antibacterial and anticancer (A549 Cell Lines) Activity. *Journal of Nanomedicine Research*, *5*(6), 1–5.

Hassan, H., Omoniyi, K. I., Okibe, F. G., Nuhu, A. A., & Echioba, E. G. (2019). Evaluation of antibacterial potential of biosynthesized plant leave extract mediated titanium oxide nanoparticles using *Hypheae thiebeace* and *Anannos seneglensis*. *Journal of Applied Sciences & Environmental Management*, *23*(10).

Hassan, S. E.-D., Salem, S. S., Fouda, A., Awad, M. A., El-Gamal, M. S., & Abdo, A. M. (2018). New approach for antimicrobial activity and bio-control of various pathogens by biosynthesized copper nanoparticles using endophytic actinomycetes. *Journal of Radiation Research and Applied Sciences*, *11*(3), 262–270.

Hussain, I., Singh, N. B., Singh, A., Singh, H., & Singh, S. C. (2016). Green synthesis of nanoparticles and its potential application. *Biotechnology Letters*, *38*(4), 545–560.

Ibrahem, K. H., Salman, J. A. S., & Ali, F. A. (2014). Effect of titanium nanoparticles biosynthesis by *Lactobacillus crispatus* on urease, hemolysin& biofilm forming by some bacteria causing recurrent uti in iraqi women. *European Scientific Journal*, *10*(9).

Jayaseelan, C., Rahuman, A. A., Roopan, S. M., Kirthi, A. V., Venkatesan, J., Kim, S. K., Iyappan, M., & Siva, C. (2013). Biological approach to synthesize TiO$_2$ nanoparticles using

Aeromonas hydrophila and its antibacterial activity. *Spectrochimica Acta Part A: Molecular and Biomolecular Spectroscopy*, *107*, 82–89.

Kalyanasundaram, S., & Prakash, M. J. (2015). Biosynthesis and characterization of titanium dioxide nanoparticles using *Pithecellobium dulce* and *Lagenaria siceraria* aqueous leaf extract and screening their free radical scavenging and antibacterial properties. *Int. Lett. Chem. Phys. Astron*, *50*, 80–95.

Khare, T., Oak, U., Shriram, V., Verma, S. K., & Kumar, V. (2019). Biologically synthesized nanomaterials and their antimicrobial potentials. In *Comprehensive Analytical Chemistry* (Vol. 87, pp. 263–289). Elsevier.

Khezerlou, A., Alizadeh-Sani, M., Azizi-Lalabadi, M., & Ehsani, A. (2018). Nanoparticles and their antimicrobial properties against pathogens including bacteria, fungi, parasites and viruses. *Microbial Pathogenesis*, *123*, 505–526.

Kubacka, A., Diez, M. S., Rojo, D., Bargiela, R., Ciordia, S., Zapico, I., Albar, J. P., Barbas, C., dos Santos, V. A. M., & Fernández-García, M. (2014). Understanding the antimicrobial mechanism of TiO_2-based nanocomposite films in a pathogenic bacterium. *Scientific Reports*, *4*(1), 1–9.

Kubacka, A., Serrano, C., Ferrer, M., Lünsdorf, H., Bielecki, P., Cerrada, M. L., Fernández-García, M., & Fernández-García, M. (2007). High-performance dual-action polymer-TiO_2 nanocomposite films via melting processing. *Nano Letters*, *7*(8), 2529–2534.

Kumar, P. S. M., Francis, A. P., & Devasena, T. (2014). Biosynthesized and chemically synthesized titania nanoparticles: Comparative analysis of antibacterial activity. *J. Environ. Nanotechnol*, *3*(3), 73–81.

Landage, K. S., Arabade, G. K., Khanna, P., & Bhongale, C. T. (2020). Biological approach to synthesize TiO_2 nanoparticles using *Staphylococcus aureus* for antibacterial and anti-biofilm applications. *J Microbiol Exp*, *8*(1), 36–43.

Liu, W., Bertrand, M., Chaneac, C., & Achouak, W. (2016). TiO_2 nanoparticles alter iron homeostasis in *Pseudomonas brassicacearum* as revealed by PrrF sRNA modulation. *Environmental Science: Nano*, *3*(6), 1473–1482.

Maurya, A., Chauhan, P., Mishra, A., & Pandey, A. K. (2012). Surface functionalization of TiO$_2$ with plant extracts and their combined antimicrobial activities against *E. faecalis* and *E. coli*. *Journal of Research Updates in Polymer Science*, *1*(1), 43–51.

Órdenes-Aenishanslins, N. A., Saona, L. A., Durán-Toro, V. M., Monrás, J. P., Bravo, D. M., & Pérez-Donoso, J. M. (2014). Use of titanium dioxide nanoparticles biosynthesized by *Bacillus mycoides* in quantum dot sensitized solar cells. *Microbial Cell Factories*, *13*(1), 1–10.

Pavlović, V. P., Vujančević, J. D., Mašković, P., Ćirković, J., Papan, J. M., Kosanović, D., Dramićanin, M. D., Petrović, P. B., Vlahović, B., & Pavlović, V. B. (2019). Structure and enhanced antimicrobial activity of mechanically activated nano TiO$_2$. *Journal of the American Ceramic Society*, *102*(12), 7735–7745.

Punitha, V. N., Vijayakumar, S., Sakthivel, B., & Praseetha, P. K. (2020). Protection of neuronal cell lines, antimicrobial and photocatalytic behaviours of eco-friendly TiO$_2$ nanoparticles. *Journal of Environmental Chemical Engineering*, *8*(5), 104343.

Rajakumar, G., Rahuman, A. A., Roopan, S. M., Khanna, V. G., Elango, G., Kamaraj, C., Zahir, A. A., & Velayutham, K. (2012). Fungus-mediated biosynthesis and characterization of TiO$_2$ nanoparticles and their activity against pathogenic bacteria. *Spectrochimica Acta Part A: Molecular and Biomolecular Spectroscopy*, *91*, 23–29.

Rajeshkumar, S. (n.d.). Novel eco-friendly synthesis of titanium oxide nanoparticles by using *Planomicrobium* sp. and its antimicrobial evaluation. *Pelagia Research Library*, ISSN: 0976-8688, 59-66.

Rao, T. N., Babji, P., Ahmad, N., Khan, R. A., Hassan, I., Shahzad, S. A., & Husain, F. M. (2019). Green synthesis and structural classification of *Acacia nilotica* mediated-silver doped titanium oxide (Ag/TiO$_2$) spherical nanoparticles: Assessment of its antimicrobial and anticancer activity. *Saudi Journal of Biological Sciences*, *26*(7), 1385–1391.

Rosi, H., & Kalyanasundaram, S. (2018). Synthesis, characterization, structural and optical properties of titanium-dioxide nanoparticles using *Glycosmis cochinchinensis* Leaf extract and its photocatalytic evaluation and antimicrobial properties. *World News of Natural Sciences*, *17*.

Santhoshkumar, T., Rahuman, A. A., Jayaseelan, C., Rajakumar, G., Marimuthu, S., Kirthi, A. V., Velayutham, K., Thomas, J., Venkatesan, J., & Kim, S.-K. (2014). Green synthesis of titanium dioxide nanoparticles using *Psidium guajava* extract and its antibacterial and antioxidant properties. *Asian Pacific Journal of Tropical Medicine*, *7*(12), 968–976.

Senarathna, U., Fernando, S. S. N., Gunasekara, T., Weerasekera, M. M., Hewageegana, H., Arachchi, N. D. H., Siriwardena, H. D., & Jayaweera, P. M. (2017). Enhanced antibacterial activity of TiO_2 nanoparticle surface modified with *Garcinia zeylanica* extract. *Chemistry Central Journal*, *11*(1), 1–8.

Shahin Lefteh, M., Sourinejad, I., & Ghasemi, Z. (2020). *Avicennia marina* mediated synthesis of TiO_2 nanoparticles: Its antibacterial potential against some aquatic pathogens. *Inorganic and Nano-Metal Chemistry*, 1–11.

Singh, P. (2016). Biosynthesis of titanium dioxide nanoparticles and their antibacterial property. *International Journal of Chemical and Molecular Engineering*, *10*(2), 275–278.

Soni, N., & Dhiman, R. C. (2020). Larvicidal and antibacterial activity of aqueous leaf extract of Peepal (*Ficus religiosa*) synthesized nanoparticles. *Parasite Epidemiology and Control*, *11*, e00166.

Subhapriya, S., & Gomathipriya, P. (2018). Green synthesis of titanium dioxide (TiO_2) nanoparticles by *Trigonella foenum-graecum* extract and its antimicrobial properties. *Microbial Pathogenesis*, *116*, 215–220.

Sundrarajan, M., Bama, K., Bhavani, M., Jegatheeswaran, S., Ambika, S., Sangili, A., Nithya, P., & Sumathi, R. (2017). Obtaining titanium dioxide nanoparticles with spherical shape and antimicrobial properties using *M. citrifolia* leaves extract by hydrothermal method.

Journal of Photochemistry and Photobiology B: Biology, 171, 117–124.

Thakur, B. K., Kumar, A., & Kumar, D. (2019). Green synthesis of titanium dioxide nanoparticles using *Azadirachta indica* leaf extract and evaluation of their antibacterial activity. *South African Journal of Botany, 124*, 223–227.

Thandapani, K., Kathiravan, M., Namasivayam, E., Padiksan, I. A., Natesan, G., Tiwari, M., Giovanni, B., & Perumal, V. (2018). Enhanced larvicidal, antibacterial, and photocatalytic efficacy of TiO_2 nanohybrids green synthesized using the aqueous leaf extract of *Parthenium hysterophorus*. *Environmental Science and Pollution Research, 25*(11), 10328–10339.

Varghese, R. J., Zikalala, N., & Oluwafemi, O. S. (2020). Green synthesis protocol on metal oxide nanoparticles using plant extracts. In *Colloidal metal oxide nanoparticles* (pp. 67–82). Elsevier.

In: Titanium Dioxide
Editor: Aparna B. Gunjal
ISBN: 978-1-68507-457-9
© 2022 Nova Science Publishers, Inc.

Chapter 5

FORENSIC APPLICATIONS OF TITANIUM DIOXIDE NANOMATERIALS

Gurvinder Singh Bumbrah[1] and Devidas S. Bhagat[2,]*

[1]Department of Forensic Science, Amity School of Applied Sciences, Amity University, Haryana, India
[2]Department of Forensic Chemistry and Toxicology, Government Institute of Forensic Science, Aurangabad (MS), India

ABSTRACT

The term "nanotechnology" was coined by U.S. engineer Eric Drexler in the 1980s. In the past few decades, nanotechnology has found exponential growth in different fields. Nanomaterials are materials that possess at least one, and usually two dimensions with less than 100nm. It can be nanoparticles, nanowires, nanorods, nanosheets, etc. These nanomaterials have gained great attention during recent years due to their unique properties including small size, stability, and low toxicity. Titanium dioxide-based nanomaterials are regularly synthesized and used for different purposes. In this chapter, recent advances in the methods of preparation of titanium dioxide

---

[*] Corresponding Author's E-mail: devidas.bhagat@gov.in.

nanomaterials are discussed along with their forensic applications in explosive detection and fingerprint imaging.

Keywords: nanotechnology; nanomaterials; titanium dioxide; forensic science; explosives; latent fingerprints

INTRODUCTION

Nanotechnology is the branch of science that deals with the study of materials having sizes in the range of 1 to 100 nm. It also includes the synthesis, characterization, and applications of nanomaterials. The science of nanotechnology has been known since the 4th century when gold was first discovered and applied as a coating agent on the Lycurgus cup developed by the Roman workers. However, in current decades, nanotechnology grows in all disciplines which facilitated mankind. It plays a vital role in biomedical science, wound healing, food services, robotics, textiles, manufacturing industries, water treatment plants, daily use products, etc. In addition to this, nanotechnology has played a potential role in the field of forensic science. However, the role of nanotechnology in the field of forensic science is new and a lot of applications in forensics sub-domains are yet to be explored. Some nanomaterials are widely used as sensors and in the device, form to examine the evidence of utmost importance to solve crimes (Salata 2004; Chen 2011; McNamara and Tofail 2017; Singh *et al.*, 2017; Pawar *et al.*, 2018; Kanodarwala *et al.*, 2019; Vijayakumar *et al.*, 2019; Prasad *et al.*, 2020).

Titanium dioxide (TiO_2) is an inexpensive, biologically, and chemically inert compound. It is widely used as a white pigment in the paint industry. It is capable of blocking the UV radiations from the sun and, therefore, is widely used in sunscreens. It is also used as a photocatalyst to remove pollutants from water (Mahmoud, Rastogi, and Kummerer, 2017).

TiO$_2$ nanomaterials are frequently used in different applications including personal care products, photovoltaic cells, biosensing, chemosensors, water purification, drug delivery, biomedical applications, etc. Due to their extensive applications, numerous studies have been conducted in the past decades dealing with synthesis, fabrication, characterization, and multiple applications of TiO$_2$ nanomaterial (Chen and Mao 2007; Chen *et al.*, 2012; Henderson and Lyubinetsky 2013; Pang *et al.*, 2013).

SYNTHESIS METHODS

TiO$_2$ nanomaterials can be synthesized in various sizes and shapes such as spherical, nanowires, nanoclusters, nanorods, nanobelts, nanotubes, and nanosheets.

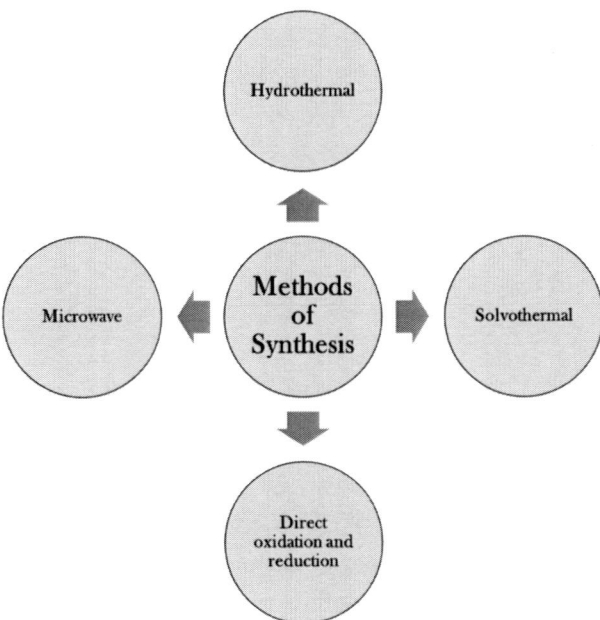

Figure 1. Common methods for the synthesis of TiO$_2$ nanomaterials.

The structural, optical, electronic, physical, chemical, mechanical and surface properties of TiO_2 nanomaterials vary with their shape and size. Nanomaterial's had a high surface-to-volume ratio that improves the sensitivity and response time. TiO_2 nanomaterials can be prepared by various methods. Figure 1 shows the most common methods used to synthesize TiO_2 nanomaterials of various shapes and sizes.

DIRECT OXIDATION METHOD

TiO_2 nanomaterials can be synthesized by oxidation of titanium metal using oxidants or under anodization. TiO_2 nanorods are produced by direct oxidation of a Ti-metal plate with H_2O_2. The crystalline phase of TiO_2 nanorods can be controlled by the addition of inorganic salts of sodium. Fluoride and sulfate salts of sodium can be used to prepare anatase form while chloride salt of sodium can be used to prepare rutile form of TiO_2. TiO_2 nanotubes can be synthesized by the anodic oxidation of titanium foil. Platinum is used as counter-electrode and crystallized TiO_2 nanotubes can be obtained after the anodized Ti plate is annealed at 500 °C for 6 h in oxygen. The potential between electrodes can be used to control the diameter and length of TiO_2 nanotubes (Wu et al., 2002a; Wu et al., 2002b; Varghese et al., 2003a; Varghese et al., 2003b; Wu et al., 2004).

HYDROTHERMAL METHOD

The hydrothermal method is performed in autoclaves under controlled temperature and/or pressure conditions. In this method, the reaction is carried out in aqueous solutions. The temperature can be increased higher than 100°C reaching the pressure of vapor saturation. The quantity of solution added to the autoclave and operated temperature controls the internal pressure produced. This method is specially used for the synthesis of small-sized TiO$_2$ nanoparticles. In addition to this, TiO$_2$ nanorods and nanowires can also be synthesized by this method (Zhang et al., 2002; Chae et al., 2003; Zhang and Gao 2003; Armstrong et al., 2005).

MICROWAVE METHOD

A dielectric material can be processed with high-frequency microwave (900-2450 MHz) radiations to synthesize the nanomaterials. In microwave conductive currents flowing in the material due to the migration of ionic components, there is transfer energy from the microwave to the material. Rapid heat transfer, selective and volumetric heating are the major merits of microwaves (Tetsushi et al., 2002; Corradi et al., 2005; Gressel-Michel et al., 2005; Ma et al., 2005; Wu et al., 2005).

SOLVOTHERMAL METHOD

The solvothermal method is a versatile method for the synthesis of nanoparticles with narrow size distribution and dispersity. This method is very similar to the hydrothermal method except that in this method non-aqueous solvent is used and is always performed at high temperature. The distributions of size and shape and crystallinity of TiO$_2$ nanoparticles synthesized by this method show more uniformity than those synthesized by the hydrothermal method. This method can be used to synthesize TiO$_2$ nanoparticles and nanorods, with narrow size distribution, with or without the need for surfactants. The crystalline phase, diameter, and length of TiO$_2$ nano-rods vary with type and concentration of precursors, surfactant, and solvent. The solvothermal method can also be used for TiO$_2$ nanowires (Kim et al., 2003a; Kim et al., 2003b; Wen et al., 2005a; Wen et al., 2005b; Wen et al., 2005c).

FORENSIC APPLICATIONS OF TiO$_2$ NANOMATERIAL'S EXPLOSIVE DETECTION

Explosive substances are important physical evidence and are frequently submitted in forensic science laboratories for their examination. Their proper examination and evaluation are useful in the investigation of criminal cases. Identification of explosives is also important to identify the terrorist group involved in the criminal activity and to prove the innocence or guilt of suspect/s. A wide range of physical and chemical methods are commonly used for the identification of explosives. The instrumental analytical techniques such as gas chromatography-mass spectrometry, high-performance liquid chromatography, Raman spectroscopy, infrared spectroscopy, UV-visible spectroscopy, neutron activation analysis, and X-ray fluorescence spectroscopy are very frequently used to identify the traces of explosives in debris samples. These methods are sensitive and specific. However, the requirement of extensive sample preparation, high-cost instruments and maintenance, interpretation of complex spectra, delayed signal response, and lack of portability are major limitations associated with these methods. Therefore, these methods cannot be used directly at the scene of the crime to detect explosives.

The advancement in nanotechnology can be of great importance in the analysis of explosive substances. Various types of nanomaterials including nanoparticles, nanowires, nanorods, etc., are used to identify explosives residues. These nanomaterials can be used in the form of sensors to detect these substances. Easy and simple preparation, low cost, high sensitivity and specificity, rapid signal response, miniaturization, and portability are some of the major merits of nanomaterial-based chemical sensors that can be used to detect traces of explosives at the crime scene in a short period.

Titanium dioxide-based modified, catalytically active, carbon-paper electrodes have been used to detect trinitrotoluene by electrochemical reduction method. The method is highly selective due to the specific electrochemical activity of the TiO$_2$/nano-Pt and TiO$_2$/nano-Au and TiO$_2$/nano-Ru composites towards the reduction of trinitrotoluene. However, TiO$_2$/nano-Ru composites show much less electrochemical activity towards the detection of TNT compared to TiO$_2$/nano-Pt and TiO$_2$/nano-Au composites. Titanium dioxide in TiO$_2$/nano-Pt and TiO$_2$/nano-Au composites plays a specific role in improving the electrocatalytic activity of platinum or gold nanoparticles and facilitating the TNT and oxygen-reduction processes. It is suggested that TiO$_2$/nano-Au and TiO$_2$/nano-Pt composites can be regarded as promising electrocatalysts for the electrochemical detection of nitro-explosives due to the inert behavior of Au and Pt to massive oxidation processes (Filanovsky et al., 2007). Boehmea et al., (2011) used a versatile titanium dioxide nanotube-based sensor for the detection of pentaerythritol tetranitrate (PETN) explosive. This device is capable of detecting PETN explosives in ultra-trace amount ~112 ppt. The reported titanium dioxide nanotube-based sensing device is easy to prepare, shows high sensitivity towards PETN, is cost-effective, and be successfully used at field or crime scenes to detect explosive materials. In another study, thin films of titanium dioxide (B) nanowires are used to detect vapors of nitroaromatic and nitrosamine explosives due to their high sensitivity and rapid response. However, the sensing response is affected in the presence of high humid conditions. The nitro groups of nitroaromatic explosives and the TiO$_2$ (B). The method is capable of detecting sub-ppb levels of nitroaromatic explosives in less than a second (Wang et al., 2011). Gokdere et al., (2019) developed TiO$_2$ based probe detection, identification, and quantification of hydrogen peroxide and triacetone peroxide explosives. 3-aminopropyl) triethoxysilane (APTES) modified-TiO$_2$NPs-based paper (APTES@TiO$_{2NPs}$) and 4-(2-pyridylazo)-resorcinol-modified-TiO2NPs-based solid (PAR@TiO$_2$NPs) sensors were developed. APTES@TiO$_2$NPs sensor forms the yellow peroxo-titanate binary complex between APTES@TiO$_2$NPs and hydrogen peroxide on chromatographic paper while PAR@TiO$_2$NPs involves the formation ternary complex between Ti(IV), PAR@TiO$_{2NPs}$ and hydrogen peroxide. LODs of PAR@TiO$_2$NPs solid sensor were 6.06×10^{-7} and 3.54×10^{-7} mol/L for hydrogen peroxide and triacetone peroxide, respectively. Although the sensitivity of a paper-based sensor is less than a solid sensor the former is easier to apply and more robust. These sensors can be used for a long time (around 7 days) without losing their stability. Xie and Liu (2019) detected and identified nitroaromatic explosives using an open-modified titanium dioxide nano-spheres-based sensor array. 5-Nitro-1,10-phenanthroline (Apne), coated on the surface of titanium dioxide nano-spheres, plays a key role in chemical recognition and absorption of nitroaromatic explosives and therefore, shows improved sensitivity towards nitroaromatic explosives. Apne broadens the absorption band edge of titanium dioxide into the visible light region. It was observed that a single TiO$_2$/Aphen sensor detects nitroaromatic explosives and improvised explosives within 8 s in a non-invasive and contactless manner.

LATENT FINGERMARKS DEVELOPMENT

Due to its light (white) color, titanium dioxide nanoparticles are frequently used to develop latent fingermarks on dark surfaces. It can be used in powder as well as in suspension form in a detergent solution. The suspension of fine titanium dioxide nanoparticles is generally used to develop latent fingermarks on wet, non-porous surfaces. However, the utility of titanium dioxide nanoparticles in developing faint and weak latent fingermarks can be enhanced by coating them with fluorescent dyes (Ramotowski, 2012; Champod, Lennard, Margot, and Stoilovic, 2016).

Saunders (1989) used titanium dioxide nanoparticles containing powder suspension to develop latent fingermarks on porous and non-porous surfaces. Wade (2002) used titanium dioxide (rutile form) nanoparticles in small particle reagent (SPR) formulation to develop latent fingermarks on dark, wet surfaces including the sticky side of black electric tape. The developed fingerprints were white and non-fluorescent.

Bergeron (2003) used the methanolic suspension of titanium dioxide nanoparticles to develop fresh and aged (7days old) bloody fingerprints on dark non-porous and semi-porous surfaces. The developed fingerprints were identifiable and second-level ridge details could be observed without background staining. Cuce et al., (2004) reported a case in which territory containing titanium dioxide nanoparticles based SPR formulation was used to develop latent fingermarks on the wet plastic bottle.

Polimeni et al., (2004) developed fresh and aged (30 days old) latent fingermarks on moist glass, painted metal, and plastic surfaces using titanium dioxide nanoparticle-based SPR. Schiemer et al., (2005) used a surfactant-containing titanium dioxide nanoparticle-based formulation to develop latent fingermarks on sticky and non-sticky sides of black electrical tape. The developed fingerprint shows second-level ridge details and can be successfully used for identification purposes. Williams and Elliott (2005) use of this formulation over Wet Wop, Wet Powder white, and adhesive side powder suspensions was recommended to develop latent fingermarks on sticky sides of adhesive tapes. In a similar report, Home Office (HOSDB, United Kingdom) recommended the use of titanium dioxide nanoparticles based wet powder suspension over Sticky-side powder for developing latent fingermarks on the sticky side of black and dark adhesive tapes (Home Office Scientific Development Branch, 2006).

Choi et al., (2007) synthesized and used fluorescent perylene diimide dye-doped titanium dioxide (TiO_2) nanoparticles to develop fresh and aged (30 days old) latent fingermarks to glass and black polyethylene. It was observed that titanium dioxide-based formulation could be used to examine third-level details in developed fingerprints on these surfaces without any background staining. The use of fluorescent perylene diimide dye-doped titanium dioxide nanoparticles over commercial fluorescent magnetic powders was suggested. The effectiveness of titanium dioxide nanoparticles towards the residues of latent fingermarks depends on the size and shape of nanoparticles; small, fine nanoparticles adhere more easily and strongly than large, coarse nanoparticles.

Reynolds et al., (2008) used and compared commercially available titanium dioxide nanoparticles of different manufacturers to develop latent fingerprints on various surfaces. It was observed that the presence, thickness, integrity, and composition of aluminum-silicon-based coats over titanium dioxide nanoparticles affect the adhesion properties and performance of titanium dioxide nanoparticles in developing latent fingerprints on various surfaces. In a similar study, Jones *et al.*, (2010) compared four different commercially available titanium dioxide nanoparticles based SPR compositions to develop latent fingermarks on the sticky side of black insulating adhesive tapes. The significant variation in the performance of alumino-silicate coated titanium dioxide-based SPR composition was observed over conventional titanium dioxide-based SPR composition due to differences in their size and morphology. It was observed that variation in morphology and chemical composition of coating material affects the surface properties of particles and interaction of fingerprint formulation with components of latent fingermark residues. These alumino-silicates are used as anti-caking agents and help maintain the adhesion and functionality of nanoparticles.

Au et al., (2011) detected and developed bloodied marks on dark, smooth, non-porous items using titanium dioxide nanoparticles based on wet powder suspension. It was observed that the application of this composition did not cause any interference in the subsequent preliminary tests for blood. It was also observed that this procedure reduces the amount of DNA recoverable from developed prints, therefore, it was suggested that this procedure should be followed only when subsequent DNA profiling from blood is not required. Despite its detrimental effects, the inclusion of this technique, as a final step, into the standard protocol for the enhancement of bloodied prints is strongly recommended by authors. This method can be used alone or in conjunction with acid dyes (acid yellow, acid violet, and acid black) to enhance bloodied marks on dark, smooth, non-porous items.

Sodhi et al., (2014) used brilliant blue R dye coated titanium dioxide nanoparticles based on fluorescent SPR formulation to develop freshly aged (up to 10 days) latent fingermarks on the wet plastic sheet, glossy magazine paper, stainless steel knife, and the sticky side of electrical tape. Peng et al., (2018) used titanium dioxide nanoparticles in powder and suspension forms to develop latent and bloody fingermarks. The developed fingerprints show second-level ridge details without any background noise. The formulation is effective in developing 180 days old bloody fingermarks and third-level ridge details can be visible in developed fingerprints. The use of titanium dioxide nanoparticles-based formulation was suggested over other commercial fingerprint powders to fresh and aged develop latent fingermarks.

CONCLUSION

In the proposed work, it has been observed that titanium dioxide-based nanomaterials are frequently used in the field of forensic science. In the case of latent fingermark imaging applications, commercially available titanium dioxide nanopowders are used either in powder or suspension form. Very limited studies have been reported in which titanium dioxide nanoparticles are synthesized and used for the development of latent fingermarks. In contrast to this, a wide range of titanium dioxide nanomaterials has been synthesized and used for the detection and identification of conventional and improvised explosives. These titanium dioxide nanomaterials are generally used as a sensor for the detection of explosives. It is noted that most of the reported work has been conducted on standard reference samples of explosive and no actual application of these titanium dioxide-based nanomaterials on the detection of explosives in real crime scene exhibits have been reported yet. Despite all these limited studies and applications, the potential utility of titanium dioxide nanomaterial's in the detection of explosives and development of fresh and aged latent fingermarks various types of dry and wet porous, semi-porous and non-porous surfaces cannot be ignored due to their ease and simple preparation, rapid signal response, high sensitivity, and specificity and cost-effectiveness. It is suggested that more studies should be conducted on the forensic applications of these titanium dioxide nanomaterials on real crime scene exhibits to solve criminal cases.

Acknowledgments

One of the author (GSB) would like to acknowledge the support provided under the DST-FIST Grant No.SR/FST/PS-I/2018/48 of Govt. of India.

References

Armstrong, A. R., Armstrong, G., Canales, J., Garcia, R., Bruce, P. G., 2005. Lithium-ion intercalation into TiO_2-B nanowires. *AdV. Mater.* 17 (7), 862-865.

Au, C., Jackson-Smith, H., Quinones, I., Jones, B. J., Daniel, B., 2011. Wet powder suspensions as an additional technique for the enhancement of bloodied marks. *Forensic Sci Int.* 204(1-3), 13-18.

Bergeron, J., 2003. Development of bloody prints on dark surfaces with titanium dioxide and methanol. *J. Forensic Ident.* 53 (2), 149-161.

Boehmea, M., Voelkleinc, F., Ensingera, W., 2011. Low-cost chemical sensor device for supersensitive pentaerythritol tetranitrate (PETN) explosives detection based on titanium dioxide nanotubes. *Sensors and Actuators B* 158, 286-291.

Chae, S. Y., Park, M. K., Lee, S. K., Kim, T. Y., Kim, S. K., Lee, W. I., 2003. Preparation of size-controlled TiO_2 nanoparticles and derivation of optically transparent photocatalytic films. *Chem. Mater.* 15 (17), 3326-3331.

Champod, C., Lennard, C., Margot, P., Stoilovic, M., 2016. *Fingerprints and Other Ridge Skin Impressions*, 2nd ed. Florida: CRC Press.

Chen, H., Nanayakkara, C. E., Grassian, V. H., 2012. Titanium dioxide photocatalysis in atmospheric chemistry. *Chem. Rev.* 112 (11), 5919-5948.

Chen, X., Mao, S. S., 2007. Titanium dioxide nanomaterial's: Synthesis, properties, modifications, and applications. *Chem. Rev.* 107 (7), 2891-2959.

Chen, Y. F., 2011. Forensic applications of nanotechnology. *J. Chin. Chem. Soc.* 58, 828-835.

Choi, M. J., Smoother, T., Martin, A. A., McDonagh, A. M., Maynard, P. J., Lennard, C., Roux, C., 2007. Fluorescent TiO_2 powders prepared using a new perylene diimide dye: Applications in latent fingermark detection. *Forensic Sci. Int.* 173(2-3), 154-160.

Corradi, A. B., Bondioli, F., Focher, B., Ferrari, A. M., Grippo, C., Mariani, E., Villa, C., 2005. Conventional and microwave-hydrothermal synthesis of TiO_2 nanopowders. *J. Am. Ceram. Soc.* 88 (9), 2639-2641.

Cuce, P., Polimeni, G., Lazzaro, A. P., De Fulvio, G., 2004. Small particle reagents technique can help to point out wet latent fingerprints. *Forensic Sci. Int.* 146S, S7-S8.

Filanovsky, B., Markovsky, B., Bourenko, T., Perkas, N., Persky, R., Gedanken, A., Aurbach, D., 2007. Carbon electrodes modified with TiO2/metal nanoparticles and their application to the detection of trinitrotoluene. *Adv. Funct. Mater.* 17, 1487-1492.

Gokdere, B., Uzer, A., Durmazel, S., Ercag, E., Apak, R., 2019. Titanium dioxide nanoparticles–based colorimetric sensors for determination of hydrogen peroxide and triacetone triperoxide (TATP). *Talanta* 202, 402-410.

Gressel-Michel, E., Chaumont, D., Stuerga, D., 2005. From a microwave flash-synthesized TiO_2 colloidal suspension to TiO_2 thin films. *J. Colloid Interface Sci.* 285 (2), 674-679.

Henderson, M. A., Lyubinetsky, I., 2013. Molecular-level insights into photo catalysis from scanning probe microscopy studies on TiO_2(110). *Chem. Rev.* 113 (6), 4428-4455.

Home Office Scientific Development Branch. (2006). Additional fingerprint development techniques for adhesive tapes. *Fingerprint Development Imaging Newsletter,* 23, 1-12.

Jones, B. J., Reynolds, A. J., Richardson, M., Sears, V. G., 2010. Nanoscale composition of commercial white powders for development of latent fingerprints on adhesives. *Sci. Justice* 50(3), 150-155.

Kanodarwala, F. K., Moret, S., Spindler, X., Lennard, C., Roux, C., 2019. Nanoparticles used for fingermark detection - A comprehensive review. Wiley Interdiscipl. *Rev.: Forensic Sci.* 1 (5), 1341.

Kim, C. S., Moon, B. K., Park, J. H., Choi, B. C., Seo, H. J., 2003a. Solvothermal synthesis of nanocrystalline TiO_2 in toluene with surfactant. *J. Cryst. Growth* 257 (3-4), 309-315.

Kim, C. S., Moon, B. K., Park, J. H., Chung, S. T., Son, S. M., 2003b. Synthesis of nanocrystalline TiO_2 in toluene by a solvothermal route. *J. Cryst. Growth* 254 (3-4), 405-410.

Ma, G., Zhao, X., Zhu, J., 2005. Microwave hydrothermal synthesis of rutile TiO_2 nanorods. *Int. J. Mod. Phys. B* 19 (15-17), 2763-2768.

Mahmoud, W. M. M., Rastogi, T., Kummerer, K., 2017. Application of titanium dioxide nanoparticles as a photo catalyst for the removal of micro pollutants such as pharmaceuticals from water. *Curr. Opin. Green Sustain. Chem.* 6, 1-10.

McNamara, K., Tofail, S. A. M., 2017. Nanoparticles in biomedical applications. *Adv. Phys. X* 2 (1), 54-88.

Pang, C. L., Lindsay, R., Thornton, G., 2013. Structure of clean and adsorbate-covered single-crystal rutile TiO_2 surfaces. *Chem. Rev.* 113 (6), 3887-3948.

Pawar, A., Thakkar, S., Misra M., 2018. A bird's eye view of nanoparticles prepared by electrospraying: advancements in drug delivery field. *J. Contr. Release* 286. 179-200.

Peng, D., Liu, X., Huang, M., Liu, R., 2018. Characterization of a novel Co_2TiO_4 nanopowder for the rapid identification of latent and blood fingerprints. *Anal. Lett.* 51 (11), 1796-1808.

Polimeni, G., Foti, B. F., Saravo, L., De Fulvio, G., 2004. A novel approach to identify the presence of fingerprints on wet surfaces. *Forensic Sci. Int.* 146S, S45-S46.

Prasad, V., Lukose, S., Agarwal, P., Prasad, L. 2020. Role of nanomaterial's for forensic investigation and latent fingerprinting – A review. *J. Forensic Sci.* 65 (1); 26-36.

Ramotowski, R., 2012. *Lee and Gaensslen's Advances in Fingerprint Technology.* 3rd ed. Florida: CRC Press.

Reynolds, A. J., Jones, B. J., Sears, V., Bowman, V., 2008. Nano-scale analysis of titanium dioxide fingerprint development powders. *J. Phys. Conf. Ser.* 126(1), 12069-12072.

Salata, O. V., 2004. Applications of nanoparticles in biology and medicine. *J. Nanobiotechnol.* 2 (2004), 3-8.

Saunders, G., 1989. Multimetal deposition technique for latent fingerprint development. In: *74th annual educational conference of the international association for identification*, Pensacola, FL, USA.

Schiemer, C., Lennard, C., Maynard, P., Roux, C., 2005. Evaluation of techniques for the detection and enhancement of latent fingermarks on black electrical tape. *J. Forensic Ident.* 55 (2), 214-238.

Singh, T., Shukla, S., Kumar, P., Wahla, V., Bajpai, V.K., Rather, I. A., 2017. Application of nanotechnology in foodservice: perception and overview. *Front. Microbiol.* 8, 1501.

Sodhi, G. S., Kapoor, S., Kumar, S., 2014. A multipurpose composition based on brilliant blue R for developing latent fingerprints on crime scene evidence. *J. Forensic Invest.* 2(2), 1-3.

Tetsushi, Y., Yuji, W., Hengbo, Y., Takao, S., Hirotaro, M., Shozo, Y., 2002. Microwave-driven polyol method for preparation of TiO_2 nanocrystallites. *Chem. Lett.* 31(10), 964-965.

Varghese, O. K., Gong, D., Paulose, M., Grimes, C. A., Dickey, E. C., 2003a. Crystallization and high-temperature structural stability of titanium oxide nanotube arrays. *J. Mater. Res.* 18, 156-165.

Varghese, O. K., Gong, D., Paulose, M., Ong, K. G., Dickey, E. C., Grimes, C. A., 2003b. Extreme changes in the electrical resistance of titania nanotubes with hydrogen exposure. *AdV. Mater.* 15 (7-8), 624-627.

Vijayakumar, V., Samal, S. K., Mohanty, S., Nayak, S. K., 2019. Recent advancements in biopolymer and metal nanoparticle-based materials in diabetic wound healing management. *Int. J. Biol. Macromol.* 122, 137-148.

Wade, D. C., 2002. Development of latent prints with titanium dioxide (TiO_2). *J. Forensic Ident.* 52(5), 551-559.

Wang, D., Chen, A., Jang, S. H., Yip, H. L., Jen, A. K. Y. 2011. Sensitivity of titania(B) nanowires to nitro aromatic and nitro amino explosives at room temperature via surface hydroxyl groups. *J. Mater. Chem.* 21, 7269-7273.

Wen, B., Liu, C., Liu, Y., 2005a. Bamboo-shaped Ag-doped TiO_2 nanowires with heterojunctions. *Inorg. Chem.* 44 (19), 6503-6505.

Wen, B., Liu, C., Liu, Y., 2005b. Depositional characteristics of metal coating on single-crystal TiO_2 nanowires. *J. Phys. Chem. B* 109 (25), 12372-12375.

Wen, B., Liu, C., Liu, Y., 2005c. Solvothermal synthesis of ultra long single-crystalline TiO_2 nanowires. *New J. Chem.* 29, 969-971.

Williams, N. H., Elliott, K. T., 2005. Development of latent prints using titanium dioxide (TiO_2) in small particle reagent, white (SPR-W) on adhesives. *J. Forensic Ident. 55(3), 292-305.*

Wu, J. M., Hayakawa, S., Tsuru, K., Osaka, A., 2002a. Porous titania films prepared from interactions of titanium with hydrogen peroxide solution. *Scripta Mater.* 46 (1), 101-106.

Wu, J. M., Hayakawa, S., Tsuru, K., Osaka, A., 2002b. Soft solution approach to prepare crystalline titania films. *Scripta Mater.* 46 (10), 705-709.

Wu, J. M., Zhang, T. W., 2004. Photo degradation of rhodamine B in water assisted by titania films prepared through a novel procedure. *J. Photochem. Photobiol. A* 162 (1), 171-177.

Wu, X., Jiang, Q. Z., Ma, Z. F., Fu, M., Shangguan, W. F., 2005. Synthesis of titania nanotubes by microwave irradiation. *Solid State Commun.* 136 (9-10), 513-517.

Xie, G., Liu, B., 2019. Fingerprinting of nitro aromatic explosives realized by aphen-functionalized titanium dioxide. *Sensors* 19 (10), 2407.

Zhang, Q., Gao, L., 2003. *Preparation of oxide nanocrystals with tunable morphologies by the moderate hydrothermal method: Insights from rutile TiO_2.* Langmuir 19 (3), 967-971.

Zhang, Y. X., Li, G. H., Jin, Y. X., Zhang, Y., Zhang, J., Zhang, L. D., 2002. Hydrothermal synthesis and photoluminescence of TiO_2 nanowires. *Chem. Phys. Lett.* 365 (3-4), 300-304.

ABOUT THE EDITOR

Aparna B. Gunjal has completed her BSc from Annasaheb Magar Mahavidyalaya, Hadapsar, MSc from Modern College Arts, Commerce and Science College, Ganeshkhind, and PhD in Environmental Sciences from Savitribai Phule Pune University, Pune, Maharashtra, India. She is working as an Assistant Professor in the Department of Microbiology at Dr. D.Y. Patil, Arts, Commerce and Science College, Pimpri, Pune, Maharashtra, India. Her research areas of expertise are solid waste management, plant growth-promoting rhizobacteria, e-waste management, bioremediation, etc. Aparna has 114 publications to her credit. She has received 6 awards for the Best Paper presentations and received travel grants. Aparna has also received Pune Municipal

Corporation Award for excellent work in Environmental Sciences Research in 2015, The Elsevier Foundation - TWAS Sustainability Visiting Expert Programme in 2018 and the Young Researcher award with Innovative Technology. She has worked on composting aspect as a Senior Researcher Assistant at Hongkong Baptist University, Hongkong. Aparna is an Editor Member of the *International Journal of Microbial Science*; *Frontiers in Environmental Microbiology*; *Journal of Microbiology and Biotechnology*, etc.

INDEX

A

absorption, 4, 21, 50, 53, 57, 59, 62, 66, 76, 108
additives, 8
adhesion properties, 110
adhesive, 109, 110, 114
adsorption technology, 24
advanced oxidation processes, 24, 39
agar diffusion, 82, 89
agglomeration, 54
agriculture, 18, 22, 71, 72
alkaloids, 84
anatase, 19, 25, 49, 50, 52, 53, 54, 55, 56, 62, 67, 104
anodization, 27, 34, 104
anthropogenic, 22
anthropogenic activities, 22
antibacterial, v, vii, 18, 20, 26, 29, 30, 34, 35, 36, 37, 39, 41, 42, 43, 44, 45, 46, 64, 68, 71, 72, 73, 77, 78, 79, 80, 81, 82, 83, 84, 85, 86, 87, 88, 89, 90, 91, 92, 93, 94, 95, 96, 98, 99
antibacterial activity, vii, 18, 35, 36, 37, 44, 45, 72, 77, 78, 79, 80, 81, 82, 83, 84, 85, 86, 87, 88, 89, 90, 94, 96, 98, 99
antibiotic, 24, 25, 79, 92
antibiotic resistance, 92
antibiotics, 18, 19, 20, 23, 25, 26, 29, 30, 35, 36, 41, 72, 78, 81
antibodies, 33
anticancer activity, 73, 85, 97
anticoagulant, 93
anti-coagulant, 79
antifouling, 9
antimicrobial, 33, 35, 39, 43, 67, 70, 72, 73, 79, 81, 85, 93, 94, 95, 96, 97, 98
antimicrobial therapy, 33
antioxidant, 73, 79, 93, 98
aphen, 108, 117
aqueous solutions, 11, 44, 104
aqueous suspension, 22
atomic oxygen, 1, 2, 5
autoclaves, 104
autoimmune disease, 33

B

bacteria, 23, 30, 35, 72, 73, 74, 75, 77, 78, 79, 80, 81, 82, 83, 85, 86, 87, 88, 89, 90, 92, 94, 95, 96, 97
bacterial cells, 80, 82, 84, 91
bacterial infection, 30, 35
bacterial pathogens, 71, 73, 75
bactericidal, 73, 77
bandgap, 21, 24, 47, 48, 49, 53, 55, 56, 57, 59, 62, 63
biochar, 25, 41
biochemistry, 39
biocompatibility, 17, 27, 32, 34, 35, 36, 84
biodegradable, 24, 74
biofilm, 29, 87, 89, 92, 95, 96
biologically, 81, 96, 102
biomedical, v, vii, 17, 18, 19, 20, 26, 32, 35, 36, 38, 39, 42, 44, 72, 102, 103, 114
biomedical applications, 26, 27, 36, 44, 103, 114
biomolecules, 32, 84
biosensing, 17, 31, 73, 103
biosensor, 18, 19, 31, 32, 38, 42, 44
biosensors, 31, 32, 44
biosynthesis, vii, 74, 75, 77, 78, 79, 80, 81, 82, 83, 84, 87, 88, 90, 92, 94, 95, 96, 97, 98
biotechnology, 39
blood, 33, 39, 40, 110, 114
bone, vii, 26, 27, 35, 38, 41, 43
breast cancer, 32, 33, 43
brilliant blue R dye, 110
brookite, 25, 49, 50

C

cable sheathing, 8
calcination, 52, 54, 56, 63, 66
cancer, vii, 18, 19, 26, 27, 31, 32, 36, 37, 38, 39, 42
cancer cells, 39
cancer therapy, vii, 18, 33
cancerous cells, 33
capping, 75, 82, 87, 88, 89
carbon dioxide, 22, 48, 49, 56
carbon nanotubes, 32
carbon-paper electrodes, 106
cell death, 80, 82, 84, 85, 90, 91
cell imaging, 19
cetyltrimethylammonium bromide, 23
chemical, 10, 18, 25, 30, 31, 34, 44, 49, 51, 55, 60, 73, 74, 75, 94, 104, 106, 108, 110, 112
chemosensors, 103
chitosan, 30, 39, 40
composition, 25, 30, 83, 110, 114, 115
compounds, 23, 33, 44, 69, 73, 77, 81, 88, 89
confocal microscopy, 89
contaminants, 8, 11, 19, 20, 21, 23, 37, 39
contamination, 1, 2, 7, 9, 11, 20, 73, 74
cost, 21, 24, 31, 37, 47, 49, 51, 54, 55, 74, 88, 106, 107, 111, 112
coumarin, 78
crystal structure, 7, 56, 58, 76
crystalline, 19, 53, 56, 62, 83, 104, 105, 116
crystallinity, 4, 55, 105
cytotoxicity, 19, 33, 37, 93

D

decolorization, 11, 68
decomposition, 9, 10, 11, 12
decontamination, 2, 20
defects, 2, 5, 22, 52, 56, 57
degradation, vii, 2, 3, 6, 7, 10, 11, 18, 19, 20, 21, 22, 23, 25, 29, 37, 38, 41, 42, 44, 46, 48, 49, 56, 57, 59, 60, 61, 62, 63, 64, 66, 67, 68, 69, 79, 85, 93, 116

Index

detection, 39, 43, 102, 107, 111, 112, 113, 114, 115
detergent, 108
diabetes mellitus, 33, 43
dielectric, 105
diffusion, 19, 31, 77, 78, 79, 81, 82, 83, 85, 86, 87, 88, 89
diffusion process, 31
disc diffusion, 77, 78, 79, 81, 82, 83, 85, 86
diseases, 19, 27, 31, 32, 33, 36, 71
DNA, 19, 39, 88, 91, 92, 110
DNA damage, 91, 92
DNA profiling, 110
dopants, 48, 50, 56, 59, 62, 63
doping, 25, 36, 48, 50, 51, 53, 56, 57, 58, 62, 63, 64, 66
doxorubicin hydrochloride, 31
drinking water, 18, 21, 22, 23
drug carriers, vii, 18
drug delivery, 19, 26, 27, 28, 29, 30, 36, 42, 43, 45, 103, 114
drug design, 46
drug release, 28, 29, 30, 38, 40, 45, 46
drug resistance, 94
drugs, 23, 26, 27, 28, 29, 30, 31, 73, 92
dye, 11, 46, 57, 63, 79, 85, 86, 93, 109, 113

E

ecofriendly, 32
electrical resistance, 115
electrocatalysts, 107
electrochemical, 24, 27, 31, 34, 38, 42, 44, 106
electrochemical oxidation, 24
electrochemical sensor, 31
electrodes, 27, 104, 106, 113
electrolytes, 27
electrometers, 73
electron, 1, 3, 4, 7, 9, 10, 11, 21, 23, 32, 48, 50, 57, 58, 88

electron beam, 1, 3
electronic structure, 50, 57, 67
electrons, 2, 5, 9, 10, 11, 49, 57, 63, 84
electrostatic, iv, 82, 84
energy, 2, 3, 5, 17, 21, 37, 48, 49, 51, 53, 55, 57, 59, 62, 64, 66, 68, 91, 105
engineering, 45, 46
environment, vii, 1, 2, 9, 10, 11, 12, 17, 22, 23, 29, 34, 35, 36, 41, 42, 49, 64, 74
environmental degradation, 13, 18
environmental management, 41, 44
enzymes, 30, 32, 33, 74, 75, 77, 82, 84, 87, 88, 91, 92
explosives, 102, 106, 107, 111, 112, 116, 117
extracts, 72, 74, 77, 79, 80, 84, 87, 92, 93, 97, 99

F

fenton process, 24
fenton reaction, 91
films, 6, 13, 14, 40, 44, 52, 65, 67, 96, 107, 112, 113, 116
fingerprint imaging, vii, 102
fingerprints, 102, 108, 109, 110, 111, 113, 114, 115
fluorescence, 33, 39, 83, 86, 106
fluorescent dyes, 108
forensic science, 101, 102, 106, 111
formation, 6, 27, 29, 56, 57, 75, 76, 82, 83, 87, 89, 95, 107
functionalization, 38, 97
fungi, 74, 75, 77, 87, 92, 96

G

gamma-ray, 8
gas chromatography-mass spectrometry, 106
genetic engineering, 19

glucose oxidase, 32
glucose sensor, 32
gold nanoparticles, 45, 107
green technology, 36
growth, 30, 31, 35, 41, 43, 52, 72, 89, 90, 92, 101
growth factor, 43

H

health, 23, 26, 35, 36, 73
hemocompatibility, 35, 38, 40, 45
high-performance liquid chromatography, 106
human, 20, 22, 23, 34, 72, 83
hyaluronic acid, 30
hyaluronidase, 30
hydrogen, 21, 31, 48, 49, 56, 57, 58, 64, 65, 66, 90, 107, 113, 116
hydrogen bonds, 31
hydrogen peroxide, 21, 90, 107, 113, 116
hydrothermal, 34, 51, 52, 55, 60, 61, 62, 98, 104, 105, 113, 114, 117
hydrothermal synthesis, 113, 114
hydroxyl, 9, 21, 78, 82, 90, 116
hydroxyl groups, 82, 116
hydroxyl radicals, 9, 21, 90

I

immobilization, vii, 18, 21
implantations, 18, 28, 34
implants, 29, 34, 35, 38, 40, 45, 46
industrial effluents, 22, 23
infrared ray, 8
infrared spectroscopy, 76, 106
inhibition, 78, 79, 81, 82, 83, 84, 88, 92
ions, 22, 29, 57, 59, 62, 63, 75, 85
iron, 42, 54, 69, 91, 96
iron transport, 91

irradiation, 1, 3, 4, 5, 6, 7, 12, 19, 21, 33, 36, 54, 64, 67, 68, 116

L

Langmuir–Hinshelwood kinetic model, 23
lanthanide metals, 59
latent fingerprints, 102, 110, 113, 114, 115
lattice, 3, 5, 53, 56, 57, 58, 59, 62, 63, 64
light, vii, 4, 5, 8, 25, 32, 33, 34, 36, 37, 41, 42, 44, 45, 46, 47, 48, 49, 55, 56, 57, 59, 60, 61, 62, 63, 64, 66, 67, 68, 69, 70, 76, 108
light scattering, 76
lipid peroxidation, 79
liquid chromatography, 106

M

mass spectrometry, 106
medical, 17, 19, 26, 29, 32, 33, 35, 72, 73
medical imaging, 73
membrane filtration, 24
methyl red, 11
microbial source, 72, 87
microscopy, 76, 80, 86, 89, 113
microspheres, 30
minimum bactericidal concentration, 78, 83
minimum inhibition concentration, 77, 83
molecular orientation, 54
molecules, 11, 31, 49, 54, 92
morphology, 30, 51, 52, 54, 63, 73, 76, 110

N

nanoclusters, 103
nanocomposites, 19, 23, 26, 28, 42, 73, 79

Index

nanomaterials, v, 18, 20, 22, 23, 25, 27, 32, 33, 35, 40, 42, 43, 44, 48, 51, 52, 69, 94, 96, 101, 102, 103, 104, 105, 106, 111
nanoparticles, v, vii, 32, 35, 36, 37, 38, 42, 43, 45, 46, 51, 53, 54, 55, 58, 62, 64, 65, 66, 68, 69, 70, 71, 72, 73, 74, 75, 76, 77, 79, 80, 81, 82, 83, 84, 85, 86, 87, 88, 89, 90, 92, 93, 94, 95, 96, 97, 98, 99, 101, 105, 106, 107, 108, 109, 110, 111, 112, 113, 114,115
nanopores, 27
nanopowders, 111, 113
nanorods, 45, 101, 103, 104, 105, 106, 114
nanoscale, 26, 33
nanosheets, 45, 101, 103
nanostructures, v, vii, 17, 47, 48, 50, 51, 52, 64
nanotechnology, 18, 43, 49, 66, 74, 75, 101, 102, 106, 113, 115
nanotube, 27, 35, 39, 40, 41, 42, 43, 44, 45, 65, 107, 115
nanotubes, 17, 27, 28, 32, 34, 35, 40, 41, 43, 44, 45, 46, 72, 89, 103, 104, 112, 116
nanowires, 17, 101, 103, 105, 106, 107, 112, 116, 117
naphthalene, 60, 63, 67
neutron activation analysis, 106
nitro aromatic, 116, 117
nitroaromatic, 107
nitro-explosives, 107
nitrosamine, 107
norfloxacin, 62, 67

O

optical properties, 4, 25, 48, 50, 54, 65, 98
orbit, 1, 2, 5, 6, 7, 8, 9
organic compounds, 69
organic matter, 9
organic pollutants, 18, 19, 22, 41, 66, 67

osteogenesis, 30, 34, 35, 45
osteosarcoma, 85
oxidants, 23, 104
oxidation, 6, 10, 21, 24, 26, 39, 59, 84, 90, 104, 107
oxidative stress, 85
oxide nanoparticles, 42, 72, 73, 94, 95, 97, 99
oxygen, 1, 2, 4, 10, 11, 21, 30, 38, 49, 57, 58, 59, 62, 91, 104, 107

P

pathogenesis, 40, 92
pathogens, 71, 73, 75, 77, 78, 79, 80, 81, 82, 83, 85, 89, 95, 96, 98
peptides, 30, 46
peptidoglycan, 82, 90
perylene diimid, 109, 113
pesticides, 18, 19, 20, 21, 22, 23, 26, 35, 36, 37, 41, 42, 67
pH, 23, 25, 28, 29, 42, 45, 51
pharmaceutical, 18, 44, 66, 69
pho regulon, 92, 95
phosphate, 91
photo degradation, 116
photocatalysis, 10, 19, 20, 21, 24, 35, 42, 44, 45, 46, 49, 50, 56, 64, 65, 66, 67, 113
photocatalyst, 1, 7, 18, 19, 20, 21, 22, 23, 24, 25, 36, 37, 41, 47, 48, 50, 52, 57, 58, 62, 63, 65, 66, 67, 69, 70, 73, 92, 102
photocatalytic, vii, 1, 2, 4, 5, 7, 9, 11, 12, 18, 19, 21, 22, 24, 25, 26, 33, 35, 36, 37, 38, 41, 42, 43, 44, 46, 47, 48, 49, 50, 51, 54, 56, 58, 59, 62, 63, 64, 65, 66, 67, 68, 69, 70, 72, 73, 94, 97, 98, 99, 112
photocatalytic efficiency, 48, 49, 52, 56, 58, 63
photodynamic therapy, 18, 19, 32, 33, 39, 41, 42, 43, 46

photoluminescence, 66, 117
photoreduction, 44, 65
photosensitizer, 33
photovoltaic, 103
physicochemical characteristics, 64
physicochemical properties, 26, 48, 51
pigment, 102
plant extract, 72, 74, 75, 77, 78, 79, 86, 87, 92, 97, 99
pneumonia, 79, 82, 83, 84, 86, 87, 89
pollutants, vii, 11, 18, 19, 20, 22, 25, 41, 44, 48, 56, 66, 67, 69, 102, 114
poly (lactic-co-glycolic acid), 28
polyethylene glycol, 28
polymer, 6, 28, 29, 30, 39, 40, 45, 96
polysiloxane–polyimide, 6
precipitation, 51, 54, 60, 61
precursors, 51, 52, 54, 55, 62, 77, 89, 105
preparation, iv, vii, 50, 51, 52, 101, 106, 111, 115
Pseudomonas aeruginosa, 36, 72, 78, 79, 80, 81, 82, 83, 85, 87, 88, 94, 95

R

radiation, 2, 4, 6, 38, 40, 55, 95
reactive oxygen, 19, 21, 25, 33, 36, 84
reactive oxygen species, 19, 21, 25, 33, 36, 84, 90
recombination, 4, 50, 54, 57, 58, 63
recombination rate, 50, 57, 58, 63
remediation, v, vii, 17, 18, 19, 20, 26, 36, 68
renewable, 23
resazurin, 85, 86
resistance, 4, 5, 6, 12, 23, 34, 49, 71, 73
response, 38, 53, 104, 106, 107, 111
rheumatoid arthritis, 33, 46
rhodamine B, 62, 116
rutile, 19, 25, 49, 50, 53, 56, 65, 67, 104, 108, 114, 117

S

sand bath method, 55
saponins, 78
satellites, 9
scanning electron microscopy, 76
semiconductor, 18, 19, 21, 24, 48, 66, 68, 70, 73
sensitivity, 9, 32, 48, 64, 90, 104, 106, 107, 111
sensors, 9, 32, 44, 73, 102, 106, 107, 112, 113, 117
siderophores, 91
signaling, 92
silica gel, 23
silicone rubber, 5
sol-gel, 13, 22, 51, 58, 60, 61, 62, 63, 68
sol-gel method, 22, 51, 58, 60, 62, 63, 68
solution, 1, 19, 34, 38, 51, 53, 55, 69, 92, 104, 108, 116
solvents, 52, 79
solvothermal, 51, 52, 61, 65, 69, 105, 114, 116
solvothermal method, 52, 65, 69, 105
solvothermal synthesis, 69
space, v, vii, 1, 2, 4, 5, 9, 12, 14, 15
space environment, 1, 2, 9
spacecraft, 1, 2, 4, 5, 6, 7, 8, 9, 11, 15
species, 10, 19, 21, 24, 25, 33, 36, 49, 51, 58, 63, 84
spectra, 4, 5, 52, 76, 106
stability, 18, 19, 28, 32, 34, 35, 36, 47, 48, 73, 101, 108, 115
structural defects, 56, 63
structure, 10, 34, 50, 52, 54, 56, 72, 76, 78, 81, 90
sunscreens, 102
superoxide, 21, 49, 84, 91
surface area, 19, 26, 30, 49, 51, 53, 54, 55, 56, 58, 59, 62, 63, 72, 85
surface chemistry, 35

surface energy, 25
surface modification, 24, 32, 34
surface properties, 51, 56, 104, 110
surfactant, 105, 109, 114
sustainable, 24, 26, 28, 30, 41, 45, 55, 74
synthesis, viii, 36, 38, 42, 48, 50, 51, 52, 54, 55, 59, 64, 65, 73, 74, 75, 76, 77, 80, 83, 84, 88, 89, 93, 94, 95, 97, 98, 99, 102, 103, 105, 114, 116, 117

T

techniques, 20, 27, 51, 76, 106, 114, 115
temperature, 4, 51, 52, 55, 69, 104, 105, 115
terpenoids, 77, 80
thermal decomposition, 51
thermal stability, 6
thermal treatment, 51
thermo-optical, 4
TiO_2, 1, 2, 3, 4, 5, 6, 7, 9, 10, 11, 12, 17, 18, 19, 20, 21, 22, 23, 24, 25, 26, 27, 28, 30, 31, 32, 33, 34, 35, 36, 37, 38, 39, 40, 41, 42, 43, 44, 45, 46, 47, 48, 49, 50, 51, 52, 54, 55, 56, 57, 58, 59, 60, 61, 62, 63, 64, 65, 66, 67, 68, 69, 70, 72, 73, 74, 75, 76, 77, 78, 79, 80, 81, 82, 83, 84, 85, 86, 87, 88, 89, 90, 91, 92, 93, 94, 95, 96, 97, 98, 99, 102, 103, 104, 105, 106, 107, 109, 112, 113, 114, 115, 116, 117
TiO_2 nanoparticles, 37, 42, 46, 51, 53, 55, 58, 62, 64, 65, 66, 68, 69, 70, 72, 75, 77, 93, 94, 95, 96, 97, 98, 105, 112
titania, 42, 45, 49, 50, 52, 53, 56, 57, 65, 96, 116
titanium dioxide, v, vii, 1, 2, 17, 18, 27, 37, 38, 39, 40, 47, 48, 49, 56, 57, 58, 64, 67, 69, 71, 72, 73, 86, 92, 93, 94, 96, 97, 98, 99, 101, 102, 106, 108, 109, 110, 111, 112, 113, 114, 115, 116, 117
titanium tetraisopropoxide, 51, 55

toxicity, 18, 24, 33, 36, 40, 41, 67, 73, 74, 77, 84, 92, 101
transmission electron microscopy, 76
transmittance, 4, 7
treatment, vii, 18, 19, 20, 22, 23, 24, 25, 26, 27, 29, 33, 34, 36, 37, 38, 43, 51, 53, 66, 71, 84, 91, 102
triacetone peroxide, 107

U

ultrasonication, 53
ultraviolet ray, 6, 12
UV light, 9, 19, 21, 24, 35, 37
UV radiation, 38, 102
UV-visible spectroscopy, 76, 106

V

viruses, 96
visible light, v, vii, 4, 25, 32, 34, 44, 45, 47, 48, 50, 55, 57, 59, 60, 61, 62, 64, 67, 68, 69, 70, 108
visible light catalyst, 48
visible-light, 4, 42, 47, 56, 62, 63, 64, 66, 69
volatile organic compounds, 59

W

waste management, 119
wastewater, vii, 18, 19, 20, 22, 23, 25, 37, 45, 64
wastewater treatment, 18, 19, 20, 22, 23, 24, 37
water purification, 18, 45, 72, 103
water resources, 20
water splitting, 48, 56, 66
wavelengths, 7, 8, 50, 53, 59
well diffusion, 78, 79, 83, 86, 87

wound healing, 102, 116

X

X-ray diffraction (XRD), 76
X-ray fluorescence spectroscopy, 106

X-ray photoelectron spectroscopy (XPS), 4

Z

zeta potential, 76